Jon –

I hope you thoroughly enjoy reading and sharing this book! Best wishes always,

Steve Fuller

Sept 2013

# Dignity

# DIGNITY

## THE ESSENTIAL
## ROLE IT PLAYS IN
## RESOLVING CONFLICT

DONNA HICKS, Ph.D.

Foreword by Archbishop Emeritus
DESMOND TUTU

Yale
UNIVERSITY PRESS

New Haven and London

Published with assistance from the Mary Cady Tew
Memorial Fund.

Yale University Press books may be purchased in quantity
for educational, business, or promotional use.
For information, please e-mail sales.press@yale.edu (U.S. office)
or sales@yaleup.co.uk (U.K. office).

Designed by Nancy Ovedovitz
Set in Adobe Garamond and HFJ Gotham type by
Duke & Company, Devon, Pennsylvania.
Printed in the United States of America by The Maple Press,
York, Pennsylvania.

Library of Congress Cataloging-in-Publication Data
Hicks, Donna.
Dignity : the essential role it plays in resolving conflict / Donna
Hicks ; foreword by Archbishop Emeritus Desmond Tutu.
p. cm.
Includes bibliographical references and index.
ISBN 978-0-300-16392-6 (alk. paper)
1. Human rights. 2. Dignity. 3. Social rights. 4. Social conflict.
5. Reconciliation. I. Tutu, Desmond. II. Title.
JC591.H53 2011
305—dc22
2011010975

A catalogue record for this book is available from the
British Library.

This paper meets the requirements of ANSI/NISO Z39.48-1992
(Permanence of Paper).

10  9  8  7  6  5  4  3  2  1

What does dignity feel like?

# Contents

ONE

## THE TEN ESSENTIAL ELEMENTS OF DIGNITY

25

# CONTENTS

# Foreword

I have come to appreciate Donna Hicks's excellent work in the field of human dignity and to value her friendship. I encouraged her to share her insights with a wider audience. This she has now done, and I congratulate her for bringing so clearly to the fore in this compelling book the concept of dignity, that inalienable God-given right of all humankind. This book is timely. We seem somehow to have forgotten that all beings are equal in dignity, the tenet in the first article of the Universal Declaration of Human Rights. The prophet in Donna Hicks brings us back to that call. She has the gift, perhaps it is her vocation, of opening to our sight a world where those most basic of human needs—appreciation, recognition, and the feeling of inherent worth—may be attained by all.

Donna Hicks tells of the time when we were together in Northern Ireland, facilitating meetings between victims and perpetrators of violence in the recent unhappy conflict in that land. Day after day, we heard the retold stories of violence and anger and the aching loss of loved ones. It was almost always the loss of dignity that drove the perpetrators to the awful acts they had committed. It was dignity regained that enabled them to face their victims. And it was dignity—the perception of worth in the other—that made reconciliation possible. I could not but reflect there on my and others' experience of apartheid in South Africa. In those dark days it was in the consciousness of our own worth and the knowledge that right must prevail and evil be overcome that our dignity sustained us. It was our sense of dignity that brought us to democracy in peaceful transition.

Dignity not only sustains but also energizes and enables. It accomplishes great things. It lifts the fallen and restores the broken. When the recognition of the good in the other is shared, it is the sense of personal dignity given that can bring peace to situations of potential conflict. People's awareness of their own dignity, their sense of worth, is the only answer to the inertia of an everyday life ruled by feelings of uselessness. How wonderful it would be if we all, every one of us, became agents of dignity, purveyors of the truth that this God-given dignity is the birthright of all.

This book is no quick and easy guide to dignity. Suggestions and challenges abound, and the guidelines that will enable us to foster good relationships are clear. But far more significant than any advice given is the driving awareness that permeates this book—that in the concept of human dignity we have in our hands, as it were, the key to the conundrum of the ages: How can peace on earth be found? Donna Hicks threads the answer clearly through her tale. God gave each of us inherent worth and value; accept it in yourself, discover and encourage it in others, and peace may just be possible.

We all long that this may become a reality in our aching world.

*Archbishop Emeritus Desmond Tutu*

# Preface

Treat people as they want to be and you help them become
what they are capable of being.

JOHANN WOLFGANG VON GOETHE

When I first considered writing a book about dignity, I thought it
would showcase the role that dignity plays in international affairs. As
a conflict-resolution specialist, I have worked as a third-party facilitator in some of the world's most intractable conflicts: Israel-Palestine,
Sri Lanka, Colombia, United States–Cuba, Northern Ireland, and
others. One experience, early in my career, put dignity in the forefront
of my consciousness. It happened in 1993, when I spent the summer
in Cambodia.

The United Nations Transitional Authority in Cambodia had set
up shop to help develop the social, legal, and political infrastructure of
the country after the genocide that had ravaged the Khmer people. I
was involved with a project, developed by Shulamuth Koenig, founding president of the People's Movement for Human Rights Learning,
that was designed to inform Khmer women of their basic human
rights under the new Cambodian constitution.

I learned a lot about dignity during my time there, but perhaps
even more about indignity. Some of the stories I heard from the

women about their experiences under the Pol Pot regime were heart-breaking. I loved introducing them to their human rights; they learned the Universal Declaration of Human Rights, proclaimed by the United Nations in 1948, and the Convention on the Elimination of Discrimination Against Women (CEDAW) inside and out. We spent a lot of time discussing the preamble of the Universal Declaration, which opens with these words: "Whereas recognition of the inherent dignity and of the equal and inalienable rights of all members of the human family is the foundation of freedom, justice, and peace in the world . . ."

I remember wondering, What would it be like if our inherent dignity were recognized on a daily basis? Would it more than fulfill the principles stated in CEDAW and the Universal Declaration? What about the way we treat one another in everyday interactions?

Hearing the painful stories of the Khmer women and witnessing their joy when they learned their rights under the new Cambodian constitution left an indelible impression on me. I felt the power behind deep listening and observed the powerful effects of seeing, hearing, and acknowledging others for what they had suffered. I realized that extending care and attention to those who had endured unspeakable atrocities helped them recover their sense of worth. Offering care and attention is, I now believe, at the heart of treating people with dignity. I saw that if indignity tears us apart, dignity can put us back together again. Dignity became the lens through which I made sense of the world from that point on.

It is not surprising, then, that I came to understand the traumatic and emotional experiences of war as assaults on people's dignity. But what I soon discovered was that the indignities that people endured were not just those spelled out in the Universal Declaration and the other U.N. conventions. Other forms of dignity violations were not mentioned in those profound documents. What about the psychological ways in which people experience wounds to their dignity? What about being excluded, misunderstood, treated unfairly, dismissed, or judged as inferior on the basis of an aspect of their identity that they could do nothing about?

After my work in Cambodia, I came back to Harvard University as deputy director of the Program on International Conflict Analysis and Resolution at the Weatherhead Center for International Affairs. There, as I continued setting up dialogues with warring parties, the issues of dignity and indignity remained alive in my mind. It never ceased to amaze me that even though the people who participated in our discussions were highly intelligent, they were ultimately unable to find a way to end the bitter struggles that were devastating their communities. Something else was going on that prevented them from resolving their differences, from putting the past to rest.

As a psychologist, I naturally gravitated toward the unspoken conversations that were taking place at the negotiation table—or perhaps *under* the table. There was always an emotional undercurrent that paralleled the discussions of the political issues, a force so powerful that it could derail productive problem-solving in a fraction of a second. Emotional riptides wreaked havoc on the people and the dialogue process. I eventually concluded that the force behind their reactions was the result of primal insults to dignity.

I suspected that nameless and unvoiced indignities were a missing link in our understanding of what keeps conflicts alive. People have a difficult time letting go of being treated badly. And if indignities are not directly named, acknowledged, and redressed—which they rarely are at the negotiating table—they take on an invisible energy of their own, showing up in the form of obstacles to a fair and equitable agreement. People need recognition for what they have suffered. And dignity violations, as ubiquitous as they are, have not been adequately recognized as a source of human suffering.

Although I started this book with my conflict-resolution community in mind, it soon became apparent that it could be a useful guide for others, in the business world, organizations, schools, and families—for anyone interested in improving the quality of his or her life and relationships.

Investigating what has been written on the subject of dignity and the role it plays in the breakdown of relationships brought me shockingly little information that was new. Some scholars have written exhaustively on the role of humiliation in international conflict, and others have inaccurately linked dignity with respect, but an extensive analysis of the issue of dignity, aimed at general readers, was nowhere to be found.

My task was made all the more complicated by the lack of a strong theoretical base in the literature from which to begin my research. Because I could not find a theoretical base, I had to build my own. A variety of disciplines contributed to the development of what I call the dignity model, an approach to understanding the primal role that dignity plays in our individual lives and in our relationships. I drew ideas and insights from evolutionary psychology, the work of William James, Immanuel Kant, and other philosophers, social neuroscience, psychology, the literature on trauma and recovery, and the field of conflict resolution. The result is a way of understanding and explaining why dignity feels so important to us and why we react so strongly when it is violated.

I have designed this book to reflect the three significant parts of the dignity model. The first part of the book introduces the essential elements of dignity: ten ways to honor dignity in ourselves and others. These elements are the building blocks of functional, healthy relationships. This section has ten chapters, each illustrating one of the elements and how to put it into practice.

The second part of the book introduces the ten temptations: traps that aspects of our evolutionary legacy have set for us, putting us at risk of violating our own and others' dignity. The ten chapters in this section take each of the temptations separately, explaining what they look like and how to effectively manage them.

The third part illustrates how to use the power of dignity to re-build a broken relationship and promote reconciliation. It tells the remarkable story of how two men on either side of the conflict in Northern Ireland reconciled after one nearly killed the other. It of-fers an alternative to forgiveness to promote reconciliation, enabling the parties on either side to repair their relationship by extending dignity to each other.

Although effort is involved in learning about dignity and how to make it a way of life, there is no better return on an investment. I have introduced the model to enough people in the international political world and in the business, education, and faith communities to know that all of us are concerned about being treated well; when we are not treated well, we suffer. Learning how to be in a relationship so that both people feel that they are seen, heard, understood, included, and given the benefit of the doubt can make a weak relationship strong and a relationship that works reasonably well work even better.

The sense of well-being that a person derives from understanding the power of dignity and putting it into practice is difficult to articu-late—it has to be experienced. The benefits of knowing how to offer dignity to others and how to maintain our own dignity are not easy to compute. We know the full value when we see our own dignity reflected back in the eyes of others.

# Acknowledgments

Writing this book required more than I was capable of on my own. Countless people have played a role in it, generously offering me insights, encouragement, and support and, perhaps most valuable of all, sharing my belief in the power of dignity.

I owe a debt of gratitude to the professional writers who have helped me along the way—Rebecca Edeleson, Patti Marxsen, Lisa Tener, and Martha Murphy. I also give thanks to friends and colleagues who have read and reread multiple versions of the book: Sousan Abadian, José Maria Argueta, Susan Muzio Blake, Steven Bloomfield, Brian Butler, Carolyn Lazar Butler, Lisa Chambers, Amanda Curtin, Richard Curtin, Rebecca Dale, Wendy Denn, Paula Gutlove, Susan Hackley, Maria Hadjipavlou, Linda Hartling, Evelin Lindner, Rhoda Margesson, Susan Colin Marks, Leonel Narvaez, Dave Nicoll, Win O'Toole, Tim Phillips, Jeff Seul, and William Weisberg. If not for my two interns, Alesandra Molina and Catherine Smail, I am not sure the idea of the book would have gotten off the ground. I owe a big thanks as well to Adam Levy for his research assistance.

It would be impossible to adequately thank my agent, Colleen Mohyde, for all that she has done. For nearly three years she has stayed by my side, often holding me up when the weight of the project felt overwhelming. She never showed any doubt in me or the book. Sharon

Hogan, my freelance editor, also deserves my profound thanks. With her abundant writing skills and sensitivity to the subject matter, she helped turn the manuscript into something I could be unequivocally proud of. I extend a special thanks to Jean Thomson Black and Mary Pasti, my editors at Yale University Press, whose input while I was preparing the final manuscript was invaluable.

Let me also give special thanks to Herbert Kelman, not only for supporting my work on the book but for making it possible for me to become the professional in conflict resolution that I am today. His wisdom, guidance, and unwavering dedication to finding a way to peace have inspired me from the beginning of my career and continue to do so to this day. I am also grateful to Archbishop Emeritus Desmond Tutu for his generous support of my work. All along, he understood what I wanted to do, not only because he believed in the power of dignity but because for him, dignity is a way of life. To his colleague Dan Vaughn I will also always be grateful. I have Shulamuth Koenig, with her passion for dignity, to thank for getting me started. I thank my family—my mother, Wanda Hicks, and my sisters, Linda Hicks, Debi Cascio, Brenda Browdy, and Sherri Barbour—for their love and encouragement. And I give infinite thanks to my husband, Rick Castino, who has read and commented on every single word I have ever written about dignity; has dropped everything, more than once, to help me through difficult times; and has provided me the financial, emotional, and spiritual security that has enabled me to fulfill a lifelong desire and to do it for dignity.

# A New Model of Dignity

*Dignity is an internal state of peace that comes
with the recognition and acceptance of the value
and vulnerability of all living things.*

On a steamy tropical morning in 2003, I walked into a room full
of civilian and military leaders in a Latin American country. The
tension in the room was as oppressive as the heat outside. There was
so much hostility, the parties in conflict wouldn't look either at each
other or at me. Although the conflict I had been called to address
revolved around the inability of these key leaders to work together,
the decades of civil war the country had experienced could not help
but contribute to the tensions I now felt.

My partner, Ambassador José Maria Argueta, and I had been invited to give a "communication skills" workshop to these elite leaders
in the hope that it would mend the ravaged relationships that I could
now see so clearly firsthand.

The president of the country walked into the room. He had come
only to introduce us and intended to leave for a meeting in the capitol
shortly thereafter. "Dr. Hicks," he said, "thank you for coming to
conduct this communications workshop with my colleagues. Could
you tell us a little about what you plan to do for the next two days?"

"Mr. President," I responded, "with all due respect, I have a feeling that a communications workshop is not what is needed here. The rifts in the relationships in this room are deep. My experience with parties in conflict is that when relationships break down to this extent, both sides feel their dignity has been violated. With your permission, I would like to shift the focus of the workshop to address these deeper issues of dignity."

Looking taken aback but remaining in control, the president turned to his scheduler and said, "Cancel my meetings in the capitol. I'll be staying for this workshop."

With his agreement I was finally able to test ideas that I had been developing for years. They were based on my cross-disciplinary research and on my two decades of experience working with warring parties around the globe. But would the concept of dignity and its use in repairing relationships resonate with these powerful officials in Latin America?

At the end of the workshop, I got my answer. One of the most resistant and unapproachable generals in the room, who had refused to look me in the eye for two days, came up to me and said, "Donna, I want to thank you. Not only did you help the relationships in this room—I think you also saved my marriage."

The dignity model was born.

*The Dignity Model.* What is the dignity model? It is an approach I developed to help people understand the role that dignity plays in their lives and relationships. It is my response to what I have observed to be a missing link in our understanding of conflict: our failure to recognize how vulnerable humans are to being treated as if they didn't matter. It explains why it hurts when our dignity is violated, and it gives us the knowledge, awareness, and skills to avoid unknowingly harming others. It demonstrates how to rebuild a relationship that has broken under the weight of conflict and suggests what to do to reconcile. The model is my response to the elephant that is always in the room when relationships break down. It names the elephant: "dignity violator."

It takes effort to learn how to honor the dignity of others, which significantly enhances the experience of being in a relationship. A good relationship feels good, but one in which both parties recognize and acknowledge each other's value and worth feels even better. Without the drag that threats place on a relationship, both parties feel free to extend themselves to each other, to open up. This is the opposite of being on guard. With safety comes the freedom to welcome intimacy and genuine connection.

The model teaches an appreciation of what we are all up against as human beings in our search for dignity. We learn how to honor it in everyday interactions with loved ones as well as strangers, how to maintain our own dignity by fighting the internal forces that tempt us to act badly, and how to resolve conflicts and reconcile with people by recognizing their inherent worth.

In the end, the message of the model is quite simple: Demonstrate the care and attention for yourself and others that anything of value deserves. That is the first and only imperative. Don't miss an opportunity to exert the power you have to remind others of who they are: invaluable, priceless, and irreplaceable. Remind yourself, too.

*The Difference between Dignity and Respect.* When I tell people that I am writing a book about dignity, they often say, "That's great. It's such an important topic." I follow up by asking what dignity means to them. They usually say, "Well, you know, people just want to feel good about themselves. They want to be treated with respect." And I say, "Okay, tell me what dignity is like. Give me an example." This is when the conversation usually stops. Most of us have a gut feeling about the word *dignity,* but few of us have the language to describe it.

3

Dignity is different from respect. Dignity is a birthright. We have little trouble seeing this when a child is born; there is no question about children's value and worth. If only we could hold onto this truth about human beings as they grow into adults, if only we could continue to *feel* their value, then it would be so much easier to treat them well and keep them safe from harm. Treating others with dignity, then, becomes the baseline for our interactions. We must treat others as if they matter, as if they are worthy of care and attention.

According to Evelin Lindner, this notion of dignity—that all human beings are imbued with value and worth—emerged in Europe in reaction to the medieval Christian belief that life was full of suffering and that humans were meant to endure suffering in this lifetime.[1] The consolation offered by the church was that the situation would improve in the afterlife. But with the advent of the Renaissance in Italy, in the fourteenth century, the very notion of what it meant to be human was opened up for discussion.[2] Philosophers and humanists began challenging the traditional beliefs, setting off a long philosophical and social discussion centered on the inherent worth and dignity of all human beings.

One Enlightenment philosopher who took up the issue of human dignity was Immanuel Kant, who, writing in the eighteenth century, introduced the idea of the "categorical imperative"—a way to determine what is morally right, independent of circumstances. One of the guiding principles of right action, Kant said, is to "act in such a way that you always treat humanity, whether in your own person or in the person of any other, never simply as a means, but always at the same time as an end."[3] Kant considered suicide morally wrong because it violated the imperative to treat not just others but ourselves as beings with inherent value and worth.

According to Kant, recognizing the dignity of all human beings means that it is unethical to exploit people or treat them merely as instruments to further one's own personal goals and interests. Honoring the dignity of others has nothing to do with any of their unique qualities or accomplishments.

Although I agree that all human beings deserve to have their *humanity* respected, human beings often behave in ways that are harmful to others, making it difficult to respect them for what they have done. I make the distinction between a *person,* who deserves respect, and a person's *actions,* which may or may not deserve respect.

Claiming that everyone automatically deserves to be treated with respect is complicated by the distinction I have just pointed out, but claiming that everyone deserves to be treated with dignity is not complicated at all. We all deserve it no matter what we do. Treating people badly because they have done something wrong only perpetuates the cycle of indignity. What is worse, we violate our own dignity in the process. Others' bad behavior doesn't give us license to treat them badly in return. Their inherent value and worth need to be honored no matter what they do. But we don't have to respect them. They have to earn respect through their behavior and actions.

Earning respect means doing something that goes above and beyond the baseline right to be treated well. If we have earned respect, we have extended ourselves to others in an admirable way. Walking out of the prison on Robben Island in South Africa, after being held there as a political prisoner for twenty-seven years, Nelson Mandela announced that he had no anger toward his captors. This extraordinary act deserves respect. He earned it.

*The Evolutionary Roots of Dignity.* To fully grasp the meaning and significance of dignity, let me put the concept into a perspective that encompasses what it means to be a human being. One of the defining characteristics of humanity is that we are feeling beings. We are equipped with five senses with which we experience others and the world around us. And we can easily affect how others feel. In fact, we have a remarkable impact on one another. With the discovery of mirror neurons, scientists now know something even more remarkable: we are hardwired to feel what others are feeling without having to say a word.[4]

Other scientists have demonstrated that human connection is crucial for survival. This new evidence that biologically connects us to one another is consistent with what many scholars of human development have argued for decades: that we are more than just individual entities hardwired for individual survival, that we are social beings that grow and flourish when our relationships are intact; our survival is inextricably linked to the quality of our relationships, and our growth and development occur in the context of relationships. Indeed, Judith Jordon and Linda Hartling argue that growth-fostering relationships are a central human necessity.[5]

What seems to be of the utmost importance to humans is how we feel about who we are. We long to look good in the eyes of others, to feel good about ourselves, to be worthy of others' care and attention. We share a longing for dignity—the *feeling* of inherent value and worth. When we feel worthy, when our value is recognized, we are content. When a mutual sense of worth is recognized and honored in our relationships, we are connected. A mutual sense of worth also provides the safety necessary for both parties to extend themselves, making continued growth and development possible.

We have an inborn desire to be treated well because we are psychologically programmed to believe that our lives are dependent on it. We cannot help but react to being mistreated. Our emotional radar is set at a very low threshold for indignities. The second we sense that someone is judging us or treating us unfairly or as if we are inferior, the emotional warning signal flashes on. Research suggests that we are just as programmed to sense a threat to our dignity—to our sense of worth—as we are to a physical threat.[6]

Thus, what appears to exist side by side with the human desire for dignity is an opposing tension: our obvious vulnerability. Although we are precious and invaluable beings, our dignity can be violated very quickly, just as our lives can be ended in the blink of an eye. We are just as vulnerable to feeling unworthy as we are to feeling worthy. Because of the primacy of relationships, our sensitivity to others and the world leaves us open to injury of all sorts and, ultimately, to the possibility of death. It appears that the feeling of loss is at the heart of human vulnerability—loss of dignity, loss of connection to others, and loss of life itself.

The human experience of worth and vulnerability is fundamentally emotional; it emanates from one of the oldest parts of our brains, from what neuroscientists call the limbic system.[7] When we sense that our worth is being threatened, we are flooded with dread and shame—with destabilizing feelings that are painful and aversive. Most of us would do just about anything to avoid these dreaded feelings, which are part and parcel of an injury to dignity. When we experience harm, our self-preservation instincts are very strong, inciting feelings of humiliation, rage, and self-righteous revenge. Some humans who have experienced chronic violations of dignity have even gone to the extreme of taking their own lives to bring an end to these intolerable feelings. Others go to the opposite extreme by killing those who caused the injury.

This highly sensitive aspect of humanity—our vulnerability to being violated by others—serves a critical, though odd, function: it promotes our survival. It warns us when danger is imminent, when someone or something threatens us; it tells us to take action to eliminate the threat. Our self-protective instincts are primed for safety, making us ready to either fight or withdraw in the service of self-preservation.[8]

Our desire for dignity has ancient evolutionary roots. Evolutionary biologists know a lot about these deep drives that explain so many of our behaviors—survival behaviors that we inherited from our early ancestors.[9] These behaviors stem from the quest to survive, and this aspect of human nature propels us throughout our lives. Some call these aspects of our human nature "instincts," since they seem to automatically and unconsciously guide us toward what to seek and what to avoid.

Importantly, however, we also have the power within us to make different choices about how we react to instincts. More recently in the story of human development, another part of our brain (the neocortex) evolved, enabling us to manage our self-protective reactions.[10]

The limbic system in our brain, that one that prompts the fight-or-flight reaction and the attendant emotions, promotes survival in another way: it encourages humans to get close to one another, to connect. Frans de Waal claims that connection is part of human biology, that human beings are hardwired to connect with one another because connection helps us feel safe rather than vulnerable. Research by Shelly Taylor and her colleagues has demonstrated that women have an apparent propensity for this alternative to fight-or-flight; they call it "tend-and-befriend." It is better to face danger together, is the argument; there is strength in numbers.[11]

Just as our limbic system can quickly signal us to disconnect from a person who harms or threatens us, it can quickly flood us with feelings of love, empathy, and compassion, compelling us to connect with another person, to find comfort in that person, to feel safer and less vulnerable, more worthy.

So humans have two distinct innate ways to ensure safety and survival: through self-preservation instincts, which prime us to alienate ourselves from those who harm us, and through self-extension instincts (tend-and-befriend), which prompt us to reach out to others and find security and comfort in friendly relationships with them. The obvious question is, Which of the two survival options has dominated the human experience?

The answer? Self-preservation seems to have dominated, not self-extension, and we are experiencing a multitude of conflicts as a result, from deadly wars that are taking countless lives to battles within families, among friends, in the workplace—everywhere human beings come into contact with one another.

Evelin Lindner, author of *Gender, Humiliation, and Global Security,* has an explanation for why our instinct for self-preservation seems to have dominated our hardwired desire to connect with others. She reports that human beings were not always so fearful of one another and draws on anthropologist William Ury's conceptualization of the stages of human history to make her argument. During the first stage our hunter-gatherer ancestors coexisted relatively peacefully. Connection triumphed over disconnection. There was enough food to go around; there was no need for competition over resources. This stage constituted approximately 95 percent of human history.[12]

A major shift took place about ten thousand years ago when the rapidly growing human population faced, for the first time, a sense of limitations. Ury calls humans at this stage "complex agriculturalists," since the scarcity of resources demanded that they adapt by working the land to produce food. The adaptation created a "fixed pie" mentality, a sense that there were only so many resources to go around. Land became divided up among people, creating opportunities for stealing and raiding.

9

Lindner points out that this shift created an "us and them" mentality; people became fearful of being invaded and raided by out-groups. She describes this transition as the beginning of fear-based notions of the other, which created, for the first time, a "security dilemma." At that point, humans became each other's predators. Out of the new need for protection from people outside a given social group, hierarchical structures developed. Some humans were at the top of the social ladder, and others were at the bottom—or, in Lindner's words, "some humans turned others into tools."[13] In the service of protection, humans turned on one another, condoning humiliating behaviors that were viewed as part of the need to survive. These acceptable acts of humiliation were not questioned in the West until the Renaissance, when Europeans challenged existing beliefs about the value of humanity.

As Lindner points out, we are now transitioning into Ury's third stage of human history—the "knowledge society." We are becoming aware of our outdated acceptance of humiliating ways of ordering society, and a new human rights culture is taking shape; the value of each and every human being is being recognized. Among other things, humanity is becoming conscious of the harmful consequences of ranking human worth. In *Somebodies and Nobodies: Overcoming the Abuse of Rank,* Robert Fuller has exposed how assigning rank to individuals undermines dignity, creating a dangerous distinction between inferior and superior human beings.[14] He has helped us see that it is not acceptable to see oneself as superior or inferior to others.

Another aspect of the transition involves resurrecting and nurturing our instinct to connect with others. I want to make it clear that not all instincts are bad. Restoring our capacity for connection (what Daniel Goleman calls "primal empathy") will enable us to find the comfort and security that only a close social bond can deliver.[15]

Being treated with dignity triggers the limbic system to release those pleasant feelings of being seen, recognized, and valued—all the life-expanding experiences that come with human connection. Instead of being flooded with fear, anger, resentment, and revenge, we experience safety in a new way. After treating one another with dignity repeatedly, after having multiple reciprocal experiences of recognizing another's value and vulnerability, we will be well on our way to discovering the possibilities that lie before us. With our inner worlds free from the turmoil and uncertainty that accompany our fear of loss of dignity, we can explore a new frontier together: what it is like to feel safe enough to be vulnerable.

Thinking of ourselves as members of the human family helps us understand that we are related by what has been passed down to us in our evolutionary history. Like all families, we unfortunately have the capacity for harming as well as loving one another. The capacity to inflict psychological injuries upon one another in the form of dignity violations is hardwired, just like the need for connection. When we suffer the wounds of feeling humiliated or diminished, an overactive emotional response can have deadly consequences. Thomas J. Scheff and Suzanne M. Retzinger tell us that unacknowledged feelings of shame (created by dignity violations) are at the heart of all human conflict.[16] The wounds do not go away spontaneously. They leave an often crippling mark, and unless attention is paid to the injuries, they can linger in perpetuity, dominating an individual's or a group's identity.

I remember a conversation I had with a member of a guerrilla organization representing an ethnic minority that was fighting for independence from the majority government. I asked him why the guerrillas were able to stay in control of their territory when they were so significantly outnumbered by government forces.

He said, "It's very simple. We are fighting to protect the dignity of our people. For the government forces, it's just a job."

The instinctive self-protective behaviors that we inherited from our ancestors were ideally suited to promote survival when resources became scarce. Those instincts are not at all suited to the complex, interdependent world we currently live in. When we feel that something threatens our well-being, our default reaction—a reaction that is unconsciously triggered and usually feels out of our control—is often an overkill reaction. In *Emotional Intelligence,* Daniel Goleman describes the experience of being captured by a default reaction as being "emotionally hijacked."[17] Our self-protective instincts are so ready to respond in threatening situations that we feel as though they take us over. Emotional hijacking happens to us all. How many times have we told ourselves that we will not let someone rile us and then, in spite of our best intentions, entered into a heated argument? This is what Goleman means when he says that these reactions have the power to hijack our best selves—the part of us that wants to work things out rationally.

Most perceived threats to our well-being today are not physical and life-endangering at all. Instead, what triggers our self-protective instincts is psychological. The triggers are, by and large, threats to our dignity. With our negative judgments and demeaning criticisms we have the power to propel each other into violent action.

Dignity threats call up a reaction from our ancient emotion center as if our lives were on the line even when they are not. When activated, our instincts do not know the difference between a physical threat and a psychological threat. All they know is that we have experienced an assault and need to be ready for action—reactive, self-protective, defensive, maybe even violent action.

The key to understanding the role that dignity violations play in our lives is understanding this point: Although external conditions and the resulting threats have changed dramatically for us in the twenty-first century, our innate, self-protective reactions have not. A majority of our threats today do not come in the form of wild animals in search of a meal. Today's threats mostly come from humans inflicting psychologically hurtful dignity violations upon one another.

*The Effect of This Legacy on Relationships.* When we perceive that we are being offended or hurt by others—when someone violates our dignity—our instinctive, self-protective hardwiring tells us that what matters most is our own well-being and survival, not the survival of the relationship. When we feel that somebody is hurting us by violating our dignity, our instincts tell us to react just as intensely as our early ancestors would have: flee or fight.

Most people know what it feels like to want to end a relationship or, at the very least, to walk out the door in the middle of a heated argument with a partner. That is the flee survival response taking hold; we want to pull out of the relationship in order to protect ourselves. When our protective instincts instruct us to fight, our tendency to befriend and to connect takes a back seat. We are pulled to denigrate the other person and perhaps to seek revenge. Instinctively, then, we want to eliminate the threat by either withdrawing from the relationship or fighting back. Both options disconnect us.

We all seem to know how to belittle and criticize others. Intellectually, we know that doing so only sets up a cycle of hurtful dignity violations. But the part of us that is driven to fight or flee is hard to convince. It does not want us to pause and reflect on what happened. It doesn't care about empathy, and it is not designed for problem-solving. All it wants is to protect us from more harm. It does not care about the consequences of its actions. It cares only about eliminating the source of the injury—either by fighting or by withdrawing.

Our desire for dignity runs deep. I believe that it and our survival instincts are the most powerful human forces motivating our behavior. In some cases, as the guerrilla leader demonstrated, our desire for dignity is even stronger than our desire for survival. People risk their lives to protect their honor and that of people in their social group; wars are fought over dignity threats. This paradoxical reaction—putting one's life on the line to protect one's dignity—puts dignity ahead of survival.

There are many objective reasons why people decide to take up arms. It would be naive not to acknowledge that point. But to minimize the role that an assault to one's dignity plays in creating conflict is not only naive but dangerous. The primal desire for dignity precedes us in every human interaction. When violated, it can destroy a relationship. It can incite arguments, divorces, wars, and revolutions. Until we fully recognize and accept this aspect of what it means to be human—that a violation of our dignity feels like a threat to our survival—we will fall short in understanding conflict and what it takes to transform it to a more fruitful interaction.

In *Humankind: A Brief History,* Felipe Fernández-Armesto presents findings from the Human Genome Project about our degree of separation from some ape relatives. A stunning finding is that we share more than 98 percent of our genetic material with chimpanzees, leaving us with less than 2 percent that sets us apart from them. He argues that the boundaries between humans and our primate cousins are so blurred that we may not deserve the special status of *Homo sapiens* at all: "If we want to go on believing we are human and justify the special status we accord ourselves—if, indeed, we want to stay human through the changes we face—we had better not discard the myth (of our special status), but start trying to live up to it."[18]

How can humans demonstrate that they are animals worthy of

special mention? I believe that the perfect proof would be to show that we can live together in this world without resorting to our knee-jerk reactions to threat, that we can treat ourselves and others with the dignity we all yearn for. However, to achieve special status, we will need to advance our self-knowledge in a way that includes and acknowledges our shared evolutionary legacy and the profound vulnerabilities that it creates for us in our relationships with one another.

We might have entered the world with strong self-protective and potentially harmful instincts, but we did not enter the world with an awareness of how much we hurt one another in the course of our own defense. Awareness requires self-understanding and acceptance. It requires work.

Ultimately, even if we are hardwired to hurt one another in the service of self-protection, it is our responsibility to know and control our reactions. We can choose to override our destructive instincts and learn more dignified ways of responding to threats—ways that would not only maintain our dignity but preserve the dignity of those who threaten us. Our opponents may have been reacting to a violation of their dignity in the first place, a violation that we unconsciously perpetrated.

Evolution did not endow us with the instinctive ability to understand the consequences of our actions. We have a hard time seeing how we set into motion the destructive power of indignity. This reactive relational dynamic is fueled by ignorance—by our lack of awareness of how we affect others. Holding up the mirror and taking an honest look at what we have done requires more than instincts. We have to tap into the part of us that has the capacity to self-reflect. We have to choose to learn how to behave. We already have inherent dignity. We just need to learn how to act like it.

If we take the issue of dignity seriously and acknowledge the direct link between being violated and the activation of our self-protective instincts, we can recognize how significant a contributor that link is to conflict. Accepting the emotional vulnerability of all humans could be the first step toward learning how to manage that vulnerability. We might even see immediate effects in our ability to get along with one another.

Just as we have devised a viable set of social contracts, from legal systems to traffic regulations, we need to devise an agreed-upon set of rules of engagement based on our understanding of dignity—on our shared human vulnerabilities and the circumstances that make us likely to trigger our self-preservation instincts. By agreeing on and honoring the elements of dignity, we could ward off a lot of conflict as well as prevent a good deal of human suffering.

*The Dignity Model in Use.* Since that initial workshop in Latin America in 2003, I have introduced the dignity model to people around the world in a variety of settings. All of the participants have had something in common: they were interested in using the model to build better relationships, often in the work environment. They wanted to establish a "culture of dignity" in which everyone would be aware of how easy it is to inflict painful wounds on others' dignity. Maybe even more important, they were eager to learn how to extend dignity to one another and create an environment in which people looked forward to being together because they felt valued.

After several workshops with different groups, it became clear that a major source of anger, resentment, and bad feelings among people who had to work together could be traced back to incidents in which individuals felt that their dignity had been violated. Every group with whom I met said that the model had put a name to experiences that had made them upset, even ready to quit, but that they had not been able to adequately articulate their reasons for being upset. Once they understood the language of dignity, they felt relieved and validated. For the first time, their suffering had a name, and they could acknowledge what they had been through.

The response is the same every time I lead a workshop—with people young and old, with people from all walks of life. Dignity is a human phenomenon. Our desire for it is our highest common denominator. We all want it, seek it, and respond in the same way when others violate it. No one wants to be harmed, and we have powerful self-preservation reactions to violations. These reactions come at a great cost, however: our need for self-preservation comes at the expense of human connection. We end up alienated from one another, going about our business as if relationships did not matter. But they do matter. Our desire for connection is deep in our genes. We are living in a false state of alienation. The quality of our lives and our relationships could be vastly improved if we learned how to master the art and science of maintaining and honoring dignity.

The debut of the dignity model in Latin America was a turning point. Before my partner and I started the workshop, I thought we were taking a risk by introducing dignity as one of the causal factors that led to the breakdown in the relationship of power in that country. I also knew that hidden in the concept of dignity was a potential torrent of unaddressed emotional issues, issues that most people are not willing to admit to, much less discuss. Scheff and Retzinger's research shows that people feel ashamed of being ashamed; they often deny it rather than talk about it.[19] I worried that the workshop participants would not want to engage in a deep discussion about such delicate and volatile matters as honor and shame.

What I was not prepared for was the willingness of the participants to discuss emotional issues. Under normal circumstances, if I ask a group to talk about a time when they felt emotionally wounded, everyone present remains silent. But this time, when I framed the question in terms of "violations to their dignity," the participants were willing to talk. Everyone had a story—several stories. I realized that the language of dignity was an acceptable way to discuss psychologically painful, humiliating, and demeaning experiences.

When I introduced the essential elements of dignity, they finally had the language they needed to articulate what had happened to them and to understand why it had felt so upsetting. The approach that I took then and have refined over the past several years was to point out that issues of dignity are not unique to this person or that group, that dignity is a profoundly emotional human issue unique to our species. It transcends race, gender, ethnicity, and all the other social distinctions. It is hard to fathom that such a significant aspect of our shared humanity has received so little attention. Left to our own (uneducated) devices, we have created an epidemic of indignity worldwide—species-wide—and we need to do something about it if we are ever going to get at this root cause of human conflict.

We do not deliberately hurt each other just for the fun of it. We are often unaware of the ways we routinely and subtly violate each other's dignity. At the same time, we are not fully aware of the power we have to make people feel good by recognizing their worth. This lack of awareness comes from not being educated about dignity. Once we become aware, we can learn how to manage our emotional reactions, which often end up hurting others, and how to communicate that we value others. Although dignity is part of our human inheritance, knowing how to nurture it is not. The actions and reactions of dignity need to be learned.

This sounds simple—all we have to do is learn how to honor each other's dignity and recognize when we are violating it. How do we learn? We have to see, first, that our lack of awareness is a problem;

second, that there is a way to handle the problem; and third, that we can make the changes necessary to do the work of dignity.

The quest for dignity is as common in the boardroom as in the bedroom, as common in the international arena as in all of our daily interactions. Our emotional responses to the way people treat us are hardwired and part of our humanity, whether we like it or not. When we are treated badly, we get angry, feel humiliated, and want to get even—often without being aware of the extent to which these primal reactions are driving our behavior.[20]

We also immediately withdraw from those who do us harm, even if we physically remain together. Fearing another assault is reason enough to shut down healthy lines of communication and trust. But often people feel that they cannot afford to leave a relationship because they are dependent on it; this happens all the time in the workplace, in marriages, and in families. Even though the relationship remains, there is a cost: openness is replaced by resentment, and we lose one of the most satisfying experiences of life—the freedom to be together without fear of being judged, harmed, or humiliated. Withdrawal and fear result in people living and working together in a state of alienation. There is no intimacy, no joy of connection. At best, the people in the flawed relationship simply tolerate one another in order to make it through the day. At worst, the relationship is characterized by hostility, and both people feel justified in demeaning the other. In short, life together is miserable.

We feel injuries to our dignity at the core of our being. They are a threat to the very essence of who we are. Worse, the perpetrators get away with harming us. And the injuries usually go unattended.

There is no 911 call for when we feel that we have been humiliated, excluded, dismissed, treated unfairly, or belittled. Neuroscientists have found that a psychological injury such as being excluded stimulates the same part of the brain as a physical wound.[21] No bones are broken, no blood appears, there are no visible signs of injury. There is harm, but the harm is felt on the inside.

What exactly gets injured? Our dignity. The painful effects of the wounds to our dignity are not imaginary. They linger, often accumulating, one on top of the other, until one day we erupt in a rage or sink into depression, or we quit our job, get a divorce, or foment a revolution. Repeated violations of our dignity undermine not only our self-worth but our capacity to be in relationships with others in ways that bring out our best and their best. What do our inaction and ignorance about these psychological wounds cost us? What do the often destructive emotional reactions they trigger cost us? A great deal is at stake.

*What Is at Stake?* At the everyday level, the aftereffects of having our dignity violated—the shame and suffering that remain—affect the quality of our lives. Scheff and Retzinger point out that that unprocessed or "bypassed" shame—shame that the victims of violations do not acknowledge or discuss because it is too shameful to admit to feeling ashamed—creates a disconnect in relationships, even for those who choose to stay together.[22] We are not free to enjoy ourselves or to extend ourselves to our families and other significant people in our lives if we are too busy protecting ourselves and licking our wounds to enjoy being with them. Suffering puts our lives on hold.

On a larger scale, bypassed shame diminishes our capacity to flourish together as human beings. Even though we have developed our intellect to astonishing heights, we are, emotionally speaking, stuck in a survival mode of existence because we have not learned how to manage our primal emotional responses to violations of our dignity, nor have we learned how to explicitly honor the dignity of others. If we continue to ignore the truth and consequences of these violations, we will remain in an arrested state of emotional development, enslaved by unacknowledged aspects of who we are as human beings.

By failing to be responsible for our responses, however unconscious those responses may be, we, by default, permit our potentially destructive instincts to be in charge of our decision making. We will see more broken hearts, more broken families, and more intractable conflicts all over the world until we understand and accept the truth about the toxic emotional power that is released when we experience threats to our dignity. As we continue to overlook this powerful contributor to human conflict and suffering, we will continue our survival mode of existence. Not until we take the matter of dignity into our own hands and make conscious choices about how we manage our hardwired, emotional reactions will change be possible.

Nevertheless, there is no doubt in my mind that we are capable of overcoming this critical human challenge to our development. I have seen miracles take place when people decide to educate themselves about the power of dignity. I have witnessed remarkable reconciliations between those who had endured years of mutual distrust, who had treated each other in the most undignified ways. I have seen the relief in people's faces when I tell them that they are feeling bad because they have endured a painful violation of their dignity. I tell them that feeling bad after a dignity violation is normal. It does not mean that something is wrong with them; what happened to them was wrong.

A participant in one of my dignity workshops said that when he read the material I sent in preparation for the event, he wept. He felt that I had articulated a sensation deep inside him, which he did not know what to call. The language of dignity had helped him name and think about his inner wounds in a way that did not make him feel ashamed or vulnerable; it legitimized his suffering. As he told the group about times when he felt that his dignity was violated, he did not hold back. He felt liberated by speaking. With the language of dignity, men and women are able to discuss for the first time those painful inner wounds that have never healed, wounds that hold them back from living life in full extension.

When I present the dignity model, one of the greatest challenges I face comes when I say that we each have the capacity to be dignity violators. People have no trouble at all seeing how they have been violated, but if I suggest that they, probably unknowingly, are violators themselves, it is a hard truth to swallow.

The only way to persuade them to accept it is to try to take away the intolerable shame of committing indignities. I explain that we all have a hardwired impulse not to want to be seen as wrongdoers and an equally strong desire to want to save face when we have done wrong. Although the experience of intolerable shame can lead to violent behavior, a tolerable level of shame—a level that promotes self-reflection and a desire to change one's behavior—can lead to reconnection with those whom one has harmed, as well as personal growth. Reviewing the ways we violate the dignity of others is not comfortable, but it is that tolerable feeling of distress that helps us change our ways.[23] Many cultures go overboard with shame. Emphasizing it might be effective in the short run, but the harmful effects can last a long time and can devastate our dignity and the learning process necessary to understand dignity violations.

Another challenge to learning about the dignity model is that more than new facts and new skills are needed for better interactions with others. Although some helpful facts and skills do come out of the model, the core of the learning involves a developmental shift in our understanding—in how we make meaning and how we come to know ourselves and the world around us. Our interpretation of what happens in the world is dependent on our experience of it.[24] The model requires us to expand this egocentric point of view, to extend and expand ourselves to take into account the perspective of others.

Integration of the experience of others into our worldview may sound simple, but what we must add to the task is the need to develop not just a cognitive understanding of others' points of view but also the "feeling of what happens" to them.[25] Restoring our capacity for primal empathy—the hardwired, emotional connection that fosters openness to others and is fundamental to taking in the entirety of the experience of others—is at the core of healthy social adjustment. Developmental shifts in consciousness do not happen without it. Emotionally identifying with others is the sine qua non of this process.

In spite of these real obstacles, I have found that most people are willing and ready to do whatever they need to do to experience the better quality of life that an understanding of dignity provides. They are tired of not feeling good about themselves, tired of being in relationships that are not working, and tired of living their lives without deep meaning and purpose. They want to become what they are capable of being.

# PART ONE

# THE TEN ESSENTIAL ELEMENTS OF DIGNITY

**Acceptance of Identity** Approach people as being neither inferior nor superior to you. Give others the freedom to express their authentic selves without fear of being negatively judged. Interact without prejudice or bias, accepting the ways in which race, religion, ethnicity, gender, class, sexual orientation, age, and disability may be at the core of other people's identities. Assume that others have integrity.

**Inclusion** Make others feel that they belong, whatever the relationship—whether they are in your family, community, organization, or nation.

**Safety** Put people at ease at two levels: physically, so they feel safe from bodily harm, and psychologically, so they feel safe from being humiliated. Help them to feel free to speak without fear of retribution.

**Acknowledgment** Give people your full attention by listening, hearing, validating, and responding to their concerns, feelings, and experiences.

**Recognition** Validate others for their talents, hard work, thoughtfulness, and help. Be generous with praise, and show appreciation and gratitude to others for their contributions and ideas.

**Fairness** Treat people justly, with equality, and in an even-handed way according to agreed-on laws and rules. People feel that you have honored their dignity when you treat them without discrimination or injustice.

**Benefit of the Doubt** Treat people as trustworthy. Start with the premise that others have good motives and are acting with integrity.

**Understanding** Believe that what others think matters. Give them the chance to explain and express their points of view. Actively listen in order to understand them.

**Independence** Encourage people to act on their own behalf so that they feel in control of their lives and experience a sense of hope and possibility.

**Accountability** Take responsibility for your actions. If you have violated the dignity of another person, apologize. Make a commitment to change your hurtful behaviors.

Dignity has these ten essential elements. We have to become aware of its essential elements to understand how to honor the dignity of others. Since our lack of awareness can make us violate others' dignity, we have to learn how that can happen. We also have to develop our sensitivity to the ways others experience us. With a developed sensitivity to others' points of view, we can minimize the times when we unknowingly violate their dignity and increase our chances of communicating that we value everyone we meet.

How did I derive the essential elements of dignity? While I was facilitating dialogues between warring parties, I spent time observing the groups' dynamics. Emotional undercurrents affected the political discussions. Both parties had reactions that were not verbal. I made sense of the emotional content of the unspoken conversation by invoking the concept of dignity and putting words to the roiling emotions: "How dare you treat me so badly? Don't you see how unfair this is? Do you think I don't notice the degrading ways you are treating me? Can't you see that I'm a human being?"

Once it was clear to me that unaddressed violations of dignity played a role in keeping the parties from coming to an agreement, I started to look for the circumstances that brought on the emotional responses. I had been trained in John Burton's human-needs theory of conflict and Herbert Kelman's interactive problem-solving approach to resolving conflict, so I was already aware that all human beings have psychological needs that, if threatened, could give rise to conflict.[1] The Burton-Kelman approach focused on providing a forum for parties to discuss unmet needs, the insights from which could be fed into the political process.

Burton's original list of "ontological" needs—needs that are fueled by the force of human development—included identity, recognition, security, and belonging.[2] When I started looking for the conditions that resulted in an observable emotional reaction during dialogues, I was already sensitized to these four needs. When someone started shouting, grew red in the face, or seemed to withdraw from the conversation, I made note of the conversation that gave rise to these physical signs of dignity violations. I also began to see other triggers that produced a visceral reaction, as when someone felt misunderstood or dismissed.

Once, a participant was describing a horrible experience, and no one from the other side of the table responded. There was no expression of remorse or compassion, not to mention an apology. I realized that if I wanted a more complete understanding of the ways people felt that their dignity was violated, I had to add to Burton's initial list. What about the desire to be understood? The desire for suffering to be seen and acknowledged? The desire to feel free from domination so that a sense of hope and possibility could blossom? The desire to be given the benefit of the doubt? The desire to be apologized to when wronged?

After compiling the list, I started to research what others had written about dignity, only to be surprised by my paltry findings. Despite numerous references to dignity in the literature, no one had specifically operationalized it as a concept. What I was looking for were answers to these kinds of questions: "If I were to say that I conducted myself with dignity, what would my behavior look like?" "If I wanted to treat someone with dignity, what would I do?" "What does it look like when I violate someone's dignity or compromise my own?" In the final accounting, the ten essential elements of dignity, built on Burton's original needs, represent years of observations: I noted the conditions (the ways people treated one another) that gave rise to the same reactions in parties in conflict all over the world when they were in dialogue together.

In the course of my research, I discovered the remarkable work of Evelin Lindner in the Human Dignity and Humiliation Studies network that she developed together with Linda Hartling.[3] Lindner's book *Making Enemies: Humiliation and International Conflict* is by far the most extensively researched work available on humiliation and its role in international conflict.[4] In it she examines a subject that has rarely been discussed in international politics; in writing it she made a major contribution to understanding one of the least understood aspects of human behavior in times of war. Although she reviews, in minute detail, the destructive and insidious nature of humiliation, she does not delve as deeply into the questions about dignity to which I was seeking answers.

Peter Coleman, a social psychologist at Columbia University, has also conducted extensive research into the contribution that humiliation makes to the intractability of conflict.[5] He and his colleagues have explored how the emotional experience of humiliation operates psychologically in people who have been subjected to it in times of war and how humiliation plays a role in exacerbating and perpetuating conflict. Although this significant work improved my understanding of the psychological dynamics that are set into motion when people experience humiliation, it did not help me in my quest to operationalize dignity.

Although the ten elements of dignity were inspired by Burton's list of the human needs for identity, recognition, security, and belonging, I became increasingly uncomfortable calling them "dignity needs." I do not think that dignity is a need. It is an essential aspect of our humanity. We do not need it, because we already have it. The physical aspects of what make us human are not "needs." Do we have a "need" for a brain, a heart, a nose? No. They are a natural part of what makes us human beings. After a long discussion with me, my friend and colleague Lucy Nusseibeh came up with the idea of calling the elements of dignity exactly that: "essential elements."[6] They became ten specific, observable ways to describe the human experience of dignity.

After running my essential elements past hundreds of participants in my workshops, I felt confident that I had found some answers to my questions. While reading the list of essential elements, keep in mind that if they describe ten different ways people experience a psychological acknowledgment of their dignity, they also describe, in reverse, ten different ways people can experience an injury to their dignity—an emotional wound that harms their sense of worth and value as a human being. Because such injuries are internal, we can easily miss them. But if you look closely enough and long enough, as I did in my years of dialogues with warring parties, you will begin to make the connection between familiar behaviors and these inner wounds.

When you see a person getting angry although he is normally reserved, ask yourself, "What kind of dignity violation has he experienced?" If a woman withdraws and turns inward although she is usually outgoing and sociable, ask, "Is she feeling dismissed or unacknowledged?" If you see a man who is unable to let go of his anger long after an upsetting incident, ask, "Did the incident stir up an old dignity wound, one that still needs to be healed?" Ask yourself, too, whether being unwilling or unable to reconnect with a person who has hurt you doesn't mean that you are still protecting your wounded dignity, leaving a part of you frozen in time.

Anger and withdrawal are normal reactions to dignity violations. When we see a person who is upset, our first reaction is usually to judge him or her negatively and to distance ourselves. We don't like to be associated with an angry person, and we usually back away from a person who has been humiliated. But if we think about what happened to that person in terms of dignity—and if we view the anger as a natural and understandable reaction to being violated, just as turning red is an indication of embarrassment—we might be able to feel empathy.

What I have found most helpful about the list of ten essential elements in my practice is that it validates and names experiences. It helps make sense of why we might feel bad after an interaction.

Knowing the essential elements of dignity helps us understand that the source of the bad feeling could be a dignity violation. Just naming the feeling a "violation of my dignity" brings relief and healing. The incident and our reaction begin to make sense. We recognize that nothing is wrong with us but that something wrong happened to us.

In chapters 2–11, I give examples of what it looks like when the essential elements of dignity are either violated or honored. I draw from various contexts, ranging from violations that occur at the international level to those that take place within families and the workplace. Although I selected these examples to highlight specific elements of dignity, many of the essential elements may be involved in a single violation. While the first story is meant to illustrate the essential element of identity, it is ultimately clear that it involves the violation of all the elements of dignity at once.

# 1

## Acceptance of Identity

Approach people as being neither inferior nor superior to you.
Give others the freedom to express their authentic selves without
fear of being negatively judged. Interact without prejudice or bias,
accepting the ways in which race, religion, ethnicity, gender,
class, sexual orientation, age, and disability may be at the core of
other people's identities. Assume that others have integrity.

One beautiful October morning in Cambridge, Massachusetts, I
was at a colleague's house for a daylong meeting to discuss the future
of a project that could make a significant contribution to improving
the political situation in the Middle East. I was a newcomer to the
project, invited by the organizers, along with five other people, to
bring new ideas to the table. Those gathered that day were a diverse
group: "peace entrepreneurs" with ties to philanthropy and the business world, area specialists from Latin America and the Middle East,
a prominent member of the arts community, and a number of us
from the international negotiation and conflict-resolution community.

One of the other visitors, an Egyptian American young man, had
returned the night before from a business trip to the Middle East.
When I was introduced to him, I said, "You've got to be jet-lagged.
Are you going to make it through this meeting?"

He laughed and said, "I'm fine. This project energizes me, and I wouldn't have missed it for the world."

I took an instant liking to him, and I was looking forward to hearing his contributions to the discussion.

In 2008, when this gathering took place, we were just a month away from presidential elections in the United States. As with every other event that I had attended that year, the meeting started with a discussion about the candidates. The Egyptian American man (I will call him Rami) told us that he was inspired by and identified with Barack Obama and that he wished he could have worked on his campaign. At first I didn't understand what he was saying. He had been involved in politics in Washington, D.C., for some time and struck me as a likely leader in Obama's efforts to win the White House. Rami saw my confused expression and said, "Don't forget, I'm a Muslim. I couldn't go near Obama. If people thought he was associated with me, it would damage his image."

I was stunned. Of course he was right. In that political environment, not many years after the September 11, 2001, attack and while American soldiers were fighting in Iraq and Afghanistan, even the mention of the word *Muslim* elicited fear in some Americans and was used to justify all kinds of discriminatory and hurtful behaviors toward Muslims in the United States and the rest of the world.

At the end of our lunch break, I saw Rami in the hallway. I wanted to acknowledge how difficult it must have been not to be able to work for Obama and how dignified I thought his response was to the situation. "Rami," I said. "I just wanted to let you know that it makes me upset to hear you say that because you are Muslim, you are staying away from the Obama campaign."

"You're kind," he said. "Thank you for saying that."

"Really, I'm not being kind. It's tragic that you feel that you have to distance yourself because of your identity. It feels dangerous. I have been concerned that no one is stepping up to the plate in this election to say that there is nothing wrong with being Muslim. I realize that Obama can't do that, but I'm not hearing it even from the political pundits. It's a blatant violation of the dignity of Muslim people, not just here in the United States but all over the world."

"I appreciate your sensitivity, but there's not much we can do about it now. We just need to get him elected; then we will work on trying to heal the wounds later. You are right that it feels like a violation of my dignity. I haven't really thought about it in that way. It is very painful for me, but I know I am doing the right thing by keeping my distance."

I said, "Not only is it an assault on your identity, and you are excluded because of it, but it feels so unfair."

One of the organizers approached us and said that it was time to restart the meeting. For a few seconds we just stood there. I didn't know what else to say. Rami put his hand on my shoulder and said, "Thank you."

❈

We can think about violations of identity in terms of human development. Throughout our lives, our inner worlds are dominated by a struggle between the ontological drives to *individuate,* to become who we are, separate from all others, and to *integrate,* to remain connected, to belong, to be a part of something greater than ourselves.[1] Thus, it makes sense that an assault on one's identity and the exclusion that results from it can be emotionally devastating.

Our identities, especially those aspects of it that are out of our control—for example, that we were born a woman, a person of color, a person with a disability—are the unique expression of who we are. To be judged a lesser person because of an inherited characteristic inflicts a wound that is especially pernicious. Inherited characteristics are used to justify not only myriad harmful behaviors but the perpetrators' sense of superiority. Anything can be justified in the name of superiority, especially the moral exclusion of "inferior" beings from one's sphere of concern.[2]

The socially constructed aspects of our identity, the product of choices we have made for ourselves that define who we are, include our professional identities (for instance, as doctors, lawyers, carpenters, businesspeople, and teachers). These aspects, added to the characteristics that we were born with, contribute to the process of individuation that makes us who we are. Either the immutable or the socially constructed aspects of our identity can be the target of dignity violations. Although violations aimed at the unchangeable aspects of our identity—our race, for example—are often more heinous, both kinds of violations have an impact. What makes the injuries all the more painful is the feeling of exclusion that victims of identity violations feel. It is next to impossible to experience a wound to one's identity and not feel a sense of being marginalized, of being excluded from the perpetrator's circle of care and concern. And the feeling of injury doesn't stop there.

Rami's case is a perfect illustration. Although the major violation he suffered was to his identity, the injury to his identity set off a cascade of other violations. In fact, I would say that all of the essential elements of dignity were violated. He was excluded from being able to participate on the basis of his Muslim identity. He was not acknowledged and recognized as a significant political player, although he had been active in Washington politics for many years. It was not safe for him to be involved in the campaign because of a possible backlash, and it was grossly unfair that he could not participate. Because of the negative stereotype of Muslims, he was not given the benefit of the doubt, making him misunderstood and disempowered. His freedom was restricted, his concerns could not be responded to—no one took the time to listen to him—and finally, there was no public attempt to right the wrong. No one took responsibility for the injuries that he and other Muslims were suffering from.

Although Rami's case provides a blatant example of a violation of one's identity, an example that was very much prevalent during the election, that type of violation occurs all the time in our everyday world. Just talk to any person who feels like an outsider. Whenever people feel that they cannot be themselves, when they feel that they don't fit in or aren't safe to be who they are without fear of being treated badly or judged as inferior to others, their dignity is wounded.

Rami's case is arguably complex—the fears and ignorance that generated the kind of xenophobic reaction that he experienced cannot be disentangled from the events of 9/11. But his story is important because it demonstrates how destructive our unfettered emotional reactions can be. Rami is an exceptional man in that he did not take the assault to his identity personally. That is not to say that he didn't feel the loss. Having one's identity trampled on in this way is hurtful and humiliating. As difficult as it was for him, he knew that if he wanted to maintain his own dignity, he could not respond in a vengeful way, even if revenge could be justified. He knew that taking such an action would help neither Obama nor himself.

What is so dangerous about violations to identity is that most of us, unlike Rami, do not show such restraint. Righteous indignation can sweep over us like a tidal wave when our identity has been compromised. "How dare you treat me this way?" is often the understandable, default response. Any number of destructive behaviors can stem from it.

John Burton wrote that when the identity needs in a group of people are not being met, the people will resort to violence, if necessary, to have those needs fulfilled— that is how powerful our desire to be seen and recognized as human beings is.[3] The violence fuse is short; it can be set off with a demeaning look as readily as with an overt racist remark. We are wired to protect our identity. With those who have been repeatedly violated, the fuse is even shorter.

James Gilligan, author of *Violence,* interviewed twenty-five hundred inmates of a maximum-security prison, most of whom were incarcerated for murder.[4] When asked why they felt compelled to kill, the majority of the inmates responded, "Because I was disrespected." Violations of our identity are like gunshot wounds to the heart.

Unless we are secure in our sense of worth, we will take these assaults to our identity personally and resort to our default reaction. If we don't flee, we may seek revenge. Since we are all human and vulnerable, we can easily become perpetrators ourselves.

❄

What happens to us on the inside when we experience an assault to our dignity? A look at our way of thinking about aspects of our identity could help us understand.

In the nineteenth century, the philosopher and psychologist William James proposed that there are two parts to ourselves. He named them the "I" and the "Me."[5] He thought of the I as the continuous presence within us that has the capacity to know the other part, the Me, which is in constant engagement with the world. The Me is the feeling, experiencing, in-the-moment part of us that interfaces with others. I have adapted his distinction of the I and the Me for the dignity model to show the important role they both play in understanding what happens when we experience an assault on the core of our being. The following story illustrates the two parts in action.

The other day at a restaurant, while waiting for friends to join me, I went up to the bar and asked for an iced tea, which I paid for as soon as the bartender supplied it. When my friends arrived, they ordered drinks, too. When we were ready to leave, I looked over our check, noticed that the bartender had charged me twice for my iced tea, and pointed the error out to him. He looked at me, rolled his eyes, snatched the check and my credit card from my hand, and huffed and puffed back to the cash register to recalculate the bill. I said to myself, "That didn't feel good, nor did I deserve it . . . just another one of those everyday dignity violations." I was tempted to snap back at him, to tell him the mistake was his, not mine. It was all I could do to restrain myself. Instead, when he returned, I took the new bill, signed it, thanked him, and walked away.

It was a close call—I could easily have lashed back. My fight reaction was powerful, especially because I felt that his demeaning reaction was undeserved. "He was the one who made the mistake! He should be apologizing to me. Instead, he's acting like a jerk."— that was one part of my internal dialogue. The other part (the part that wanted to maintain my dignity) calmed the outraged me down, pointing out that by losing my cool, I would be jeopardizing my own dignity. I was caught in a struggle between two parts of myself, both equally true to who I am. My Me wanted to lash out, but my I won, saving me from violating not only the dignity of the bartender but my own.

❋

Now let's see how I have adapted James's framework to identify and name the internal battles we all fight when our dignity is hurt.

I think of the Me as the part of ourselves that can function outside our awareness. Driven by the need to be accepted by others (thanks to our evolutionary legacy), the Me seeks external validation of its worthiness. It cannot feel good about itself unless it wins praise or approval. It is vulnerable to others' judgments and criticisms and reacts to threats to its dignity by defending and protecting itself.

The Me is often distrustful of others, cognizant of their potential to do harm. When we are situated in the Me part of ourselves, our inner world is dominated by concerns about the self: "Am I good enough? How do I compare to others? Am I acceptable?" In assessing others, the Me is often judgmental and critical; it constantly looks for ways to diminish others so that it can look good itself. When someone gives us a compliment, or acknowledges that we have done a good job, Me feels good. Either way, whether we are praised or criticized, Me responds—it longs for one and dreads the other.

The Me is also the part of us that gets into conflict with others. It wants to get even when someone hurts us; it seeks revenge. The need for revenge gets us into trouble because by lashing out or getting even, we end up violating the other person's dignity as well our own—and tumbling downward into constant conflict.

The other part of the self is as true a part of us as the Me. The I knows that its significance and worth are not negotiable. It doesn't need validation from outside sources. It is the enduring aspect of who we are; its dignity is unconditional.

When we are situated in the I, there is no such thing as a good self and a bad self. The I owes its significance and worth to our being part of the human family, part of life itself. It does not need acknowledgment of its right to hold itself in esteem. The I just is. Some would say that the I is the spiritual aspect of the self, connected to everything else in the universe, that it is part of the miracle of nature and human existence.

When we are centered on the I, the world is a wonder. We are curious about everything—from ourselves and others to the multitude of mysteries around us. We are joyful, creative, expansive, and at peace. We are aware that being connected to others is our natural state. Once we are fully anchored in our I, we care about how others feel about us because we care about others, not because we need them to make us feel good.

When we are centered on the I, we do not seek praise and approval; instead, we seek to expand and to make our lives meaningful. We are unencumbered by self-doubt and free to explore ways to put our talents to good use.

The I keeps us steady when our Me is threatened or hurt. It is the part of us that we can always retreat to when our dignity has been violated. It stops our Me from wanting to get even with the person who offends us. *The I is stronger than the Me.* It has the power to resist the temptation to seek revenge and the power to maintain our dignity no matter how badly someone treats us. It knows, or can learn, that we do not want to let the bad behavior of others determine how we act, and it knows that by extending dignity to others, we strengthen our own.

The dignity model seeks to reconcile the I and the Me. Reconciling them is our first order of business. Because we are human, we cannot expect to eliminate our need for praise and approval. Admiration feels good, and we all enjoy it. Nor can we expect not to feel pain when we are hurt. We need to develop a pathway between the I and the Me so when the Me gets injured, the I can come to its rescue, protecting us from the Me's instinctive, often self-destructive impulses, even if they are in the service of self-preservation.

Until we become aware of the two parts of who we are, until we name them and reconcile them, we tend to banish the Me, driving it out of our consciousness, abandoning a part of our humanity that we would be wiser to be aware of and control than to leave to its own devices. Some psychologists argue that the integration of the competing parts of ourselves is the hallmark of a healthy person. The lack of integration creates both inner chaos and rigidity.[6]

Every time I present this material to participants in my workshops, the vast majority of them say that they locate themselves in the Me most of the time. They realize how much of their internal focus is on wanting to look good to others and comparing themselves to others. Most of us are looking for validation from each other and feeling unworthy without it. We want the kind of dignity that the I can give us, but instead we are looking for it outside ourselves.

This knowledge—that we are both the I and the Me—helps us to understand why it is so easy to get into conflict with others. Our Me's collide, and we protect ourselves, even at the cost of our own dignity.

When we have learned to embrace these two, often-competing aspects of who we are, we naturally end up, most of the time, feeling ambivalent. Even when we know that it is right to restrain ourselves rather than lash out, we still feel the desire to fight back. My injured Me wanted to verbally pummel the bartender for blaming me for his mistake. I came perilously close to creating a scene at the bar, violating his dignity and dragging myself down in the process, but fortunately my I took over.

We know that these two parts of us exist and that we will often therefore be in a state of ambivalence. Having destructive thoughts doesn't make us bad people. It simply means that there are two compelling and competing aspects of who we are, and it is often difficult to reconcile them. But with a little knowledge, we can put ourselves in a position, with a strengthened I, to make the right choices.

Acknowledging that we have the capacity to do harm when we have been harmed can not only leave us ambivalent but also keep us humble. Knowing that we ourselves could stab with knife or tongue prevents us from taking a superior attitude when we see others inflicting harm.

If we look again at the way Rami handled the assaults to his dignity, we can see that even though he felt hurt and upset about his circumstances (his Me definitely took a hit), he maintained his dignity by letting his I be in charge of his behavior. He didn't take the injury personally. He was an unusual young man who understood that his dignity was not in the hands of others; he never lost sight of his inherent worth.

# 2

## Inclusion

Make others feel that they belong, whatever the relationship—
whether they are in your family, community, organization, or nation.

After spending fifteen years in Madison, Wisconsin, having built a rich life with a strong sense of community, good friends, and an abiding love of the area's natural beauty (Madison is situated among five lakes), I left in September 1991, only days after submitting my completed dissertation to the registrar's office at the university. Leaving was difficult because I was saying good-bye to a life that was good beyond what I had ever thought possible. I had felt seen, supported, honored, and loved by my community—I belonged.

Yet a strong force was tugging at me, making me let go of the unconditional safety and inclusion that Madison represented. I had been offered a postdoctoral fellowship at Harvard University to study with a world-renowned expert in the field of international conflict, Herbert Kelman. As I pulled out of my driveway in a truck heavily packed with my belongings, I knew that I had to keep focusing straight ahead. I turned up the radio and fought the urge to look in the rearview mirror to take one last peek. I put on my sunglasses, turned to my friend who was accompanying me, and asked, "Are you sure you know where we're going?"

No one was more surprised than I was by the outcome of a call I had made to Professor Kelman earlier that spring, asking him for advice about where I should go for training in international conflict resolution. I had met him a year earlier at a conference and was captivated by his psychological approach to resolving conflict. During my years in graduate school, I had wanted to know more about the psychology of conflict, about what drives people to kill one another. At the time, very few people in the field of international relations were concerned about the psychological dimension of conflict. The prevailing sentiment was to dismiss it, and those who believed that psychology played a significant role in international affairs were often marginalized. Professor Kelman was one of a handful of people who had the temerity, as well as the intellectual gravitas, to stand up and say that what happens in our inner worlds matters.

On the phone, after laying out my options for a postdoctoral fellowship, Kelman finally said, "Well, Donna, if none of those turns out, you could always come here and do a fellowship with me. I have a group of graduate students and postdocs from all over the world who meet regularly, all of whom are studying my methodology. You would be most welcome to join us."

I was shocked at the unexpected offer. I can't remember what I said to him, but I must have said yes. He said that his assistant would send me information about the fellowship. I hung up and walked around my apartment in a daze.

Once I arrived in Cambridge and moved into my new apartment, panic set in. In all the excitement of the move, I hadn't imagined what it was going to feel like to be away from home, alone, in the heart of one of the most prestigious intellectual and cultural communities in the country. Nor had I expected the recurrence of the dreaded feeling of being "out of my class." I grew up in a rural community in upstate New York, in a family that struggled to make ends meet. "Hicks from the sticks" was what my friends jokingly called me. I was well aware of my class background, and I often felt a sense of inferiority around people who had grown up with privilege. After I left home at age seventeen, I dreaded the inevitable question, "So what does your father do?"

Here I was, out on the margin again. How could I feel as though I belonged at Harvard? The feeling of being "less than" made me want to run away. I didn't run, but I did decide to put off my trip to the psychology department to let Professor Kelman know that I had arrived. Deciding to go to a gym instead, I grabbed my gear and umbrella and walked about a mile in pouring rain to get there.

Upon my arrival, head-to-toe drenched, I asked where the locker room was. Then someone told me I needed my own lock. I stood there, dripping, loaded down with my workout bag and umbrella, and almost started to cry. I wanted Madison, my old friends, my old apartment, and the safety of knowing I belonged.

Before I could walk away, a woman grabbed my arm, obviously sensing my distress. She said, "Look, don't leave, you can share my locker. Are you here for the aerobics class at noon?" I thanked her and said yes. We made small talk while we changed our clothes. When I told her that I was starting a postdoctoral fellowship with Professor Kelman in the psychology department, she paused, stared, and said, "Are you Donna Hicks?"

In utter disbelief, I nodded my head. With a huge smile on her face, she said, "My name is Ariella Baiery, one of Herb's graduate students. We've been waiting for you!"

I never again felt that I didn't belong. I cannot adequately convey the impact of Ariella's kindness and welcome that day. My feeling of isolation, my sense of inadequacy, and my fear of having made a huge mistake started to evaporate as soon as I met her, and it continued to dissipate when I met the rest of Kelman's group of students in PICAR, the Program on International Conflict Analysis and Resolution.[1] They treated me as if I were family. And the truth is, Herbert Kelman and his wife, Rose, did the same for everyone. We all became a part of their flock and found refuge with them within the wider world of Harvard University.

Although the fears of being marginalized and excluded went away quickly because of the Kelmans' largesse and the overall decency of every member of the PICAR group, it was still daunting to be at Harvard. Being part of PICAR helped me recover my sense of belonging, but on a larger scale, the challenges remained for some time.

When I reflect back on those early days, I often wonder what it must be like for others who come to Harvard from backgrounds different from those of the white, affluent majority or who are even more clearly and distinctly different. I was able to "pass" for many reasons, not the least of which was that I looked like a white European American. I could disguise my feelings of being out of my element, the sense that I didn't belong. But what if I had been black, Latino, or a member of any other disenfranchised minority group? What would it be like for them to arrive at Harvard? If their experiences were anything like mine, it wouldn't be easy to walk into a place dominated by a majority so different from them. Fitting in is no small matter. I know people who spent years at Harvard and never felt that sense of belonging. There were no Ariellas in their experience.

I wonder, too, what it is like for immigrants coming into this country. My mother is a first-generation Polish American who didn't speak English until she started school. Her parents arrived in this country when they were sixteen years old. I know how her family suffered from that feeling of being less-than and inferior. Her parents struggled on their farm, trying to raise sixteen children with an income below the poverty line. I often wondered whether my grandparents had so many children to compensate for the loss of a sense of belonging. Were they creating their own community?

Even if immigrants coming here find good jobs and seem to be managing financially and assimilating into the community, the way they feel on the inside is another matter. Money cannot satisfy the profound human desire to belong, to feel the comfort and safety that acceptance brings. When I think of the way immigration is handled here—when I reflect on the absence of basic human decency (and dignity) not only in the discussion of immigration policy but also in the treatment of newcomers, I wonder how people in positions of power sleep at night.

Consider, for a moment, the emotional toll on immigrants who leave loved ones and all that is familiar behind to improve the quality of life for their children and their families back home. Imagine the sense of loss they experience when they arrive here, only to be met, as is often the case, with treatment as something less-than. Then imagine for a moment what it would be like for them if a bright-eyed Ariella were there to say, "Welcome, we've been waiting for you."

# 3

## Safety

Put people at ease at two levels: physically, so they feel safe from bodily harm, and psychologically, so they feel safe from being humiliated. Help them to feel free to speak without fear of retribution.

My husband, Rick, and I were invited to the home of new friends for dinner to celebrate the seventh birthday of their youngest child, Seth (not his real name). When we pulled into the driveway, we saw about a dozen children playing soccer in their big backyard. The house and gardens were beautiful—our friends had spent time and effort on landscaping—and because it was such a warm night, they had decided to have the party outside. The dinner table was set underneath a huge maple tree strung with little white lights.

Before we got out of our car, we sat for a while watching the children play. Not able to have children ourselves, we both get a little wistful at times like these. After a minute, still staring at the children, my husband said, "They have it all, don't they?"

We joined the other adults on the patio. Our hosts, Margot and Tom (not their real names), gave us a warm welcome, then introduced us to the other guests, most of whom were the parents of Seth's friends.

When Margot announced that dinner was nearly ready, Rick and

I volunteered to help bring the food outside. Seth came running into the kitchen, out of breath, unsuccessfully trying to hold back tears. Tom looked at him and said, "What's the matter? Why are you crying? You look like a baby."

Seth burst into tears and ran into his mother's arms. He told her that one of his friends had yelled at him in front of everybody because he had messed up a goal.

Tom said to Margot, "Don't baby him."

"What are you talking about? He's hurting."

Tom walked up to her and said, his face only inches from hers, "You have made him into a sissy. So what if his feelings are hurt? He needs to toughen up. He can't come running to you every time something goes wrong."

With her hands on her hips and her chest heaving, she said to Tom, "I'll tell you what's wrong; you're what's wrong. Don't you dare talk to me like that." Margot stormed out of the room.

Tom turned to Seth and said, "Get back out there with the other kids and stop being a mama's boy."

Seth left the room with his chin on his chest and with arm across his face, wiping away tears. Tom said to us with a nervous laugh, "A little shaming always works." He picked up a tray of food and headed for the patio.

Rick turned to me and said, "Never trust appearances."

❋

This story shows what a violation of dignity's essential element of safety looks like. Specifically, Seth's psychological safety was at stake. Seth experienced a double hit. He was publicly humiliated when one of his friends yelled at him, in front of everybody, for not making a goal—which he must have felt bad about already. Having attention drawn to his mistake made flubbing the goal even worse, to the point where he bolted. Flight is a typical reaction to being humiliated, as we have seen. Seth fled into the house for consolation. But he did not get the nurturing and acknowledgment he needed. Instead, he was hit again. His father called him a sissy, and he did it in front of others. Humiliating someone in front of other people can be devastating.

When we are psychologically injured, the area of our brain that is activated is the same area that is activated when we experience a physical injury, as research by Naomi Eisenberger and Matthew Lieberman has shown.[1] We wouldn't think twice about rushing our child to an emergency room if he broke his arm or leg; the pain and suffering of a physical injury is acknowledged immediately. When his spirit is broken by shame or humiliation, when damage is done to his sense of worth, there was, in Seth's case, nowhere to go to take care of the wound. Running to his mother was his best option. But being told, as Seth was, to "be tough" and "stop being a mama's boy," makes it a fairly sure bet that the internal injury will grow and fester, contaminating his sense of worth. We humans need acknowledgment for what we have suffered, and when we don't get it, the temptation to think we deserved our misfortune comes naturally. Safety and vulnerability share a complex connection.

Listening to parents yell at each other also undermines a child's sense of security, and it models undignified behavior. Sad to say, both parents felt justified in saying hurtful things to one another. Each felt the other was wrong. Our need to be right is a powerful motivation for inflicting psychological harm on another—for violating another's dignity. Righteous indignation is used to justify bad behavior all too frequently.

I once heard William Sloane Coffin, pastor of the Riverside

Church in New York, being interviewed on the radio. He said that self-righteousness was a scourge because it didn't leave room for self-criticism. In the heat of the moment, neither Tom nor Margot stopped to reflect on their own behavior. They were both overtaken by their fight instincts. Their Me's were in combat. Their reactions were so strong that the presence of other adults, who were not close friends, did not deter them.

Tom and Margot are not bad people. They have good intentions; they both want the best for their youngest child. But like so many of us, they are not aware of their blind spots—in this case, their instinctive violations of each other's dignity and the dignity of their child, especially under stressful circumstances.

What would it have looked like if Tom and Margot had handled their son's emergency with dignity? Margot was on the right track—she acknowledged how hurtful it was for Seth to be shamed in front of his friends. After the acknowledgment, she could have reminded him that he was a wonderful boy and that he shouldn't take his friend's outburst to mean that he himself had something wrong with him. His friend was upset because his team hadn't won the game, that's all. And Seth had missed a goal, which anyone could have done.

After she was sure that she had successfully helped Seth reinterpret the event, Margot could have encouraged him to go back out and play. She could have also encouraged him to go to his friend and say that he was sorry he had missed the goal—he had wanted to win the game, too.

Most people, young or old, respond positively to a comment like that. It is a way to acknowledge the experience of the other, even the one who inflicted the wound. It is a way to say that the victim, too, has a perspective on the situation, one that he is capable of seeing now that he is no longer hurting. A gesture of acknowledgment doesn't let the perpetrator off the hook, but it gives the victim a chance to regain his own wounded dignity and open up his perspective to include the experience of the other.

It would not be surprising to discover that Tom suffered crippling shame as a child. His father probably shamed him and told him to toughen up, too. This is how ignorance and the pain it causes get passed down like a dominant gene. Old childhood injuries create blind spots for us unless we get the help we need to recover from them.

Tom's blind spot was not being aware of how intolerable shaming can be to a child. He had dissociated pain from shame long ago to survive. A little education about how not to violate the dignity of children could have helped Tom, not only with his parenting but also with his self-knowledge. He could have understood the effects that extreme shame had on him when he was growing up. These early imprints of indignity—the memories of being painfully shamed early in childhood—continue to affect us and our dealings with others throughout our lives. Unless we become aware of them, heal from them, and make a conscious decision not to let them determine our actions, we continue to hurt others, jeopardizing their and our own dignity and threatening all our relationships.

❀

The early imprints of dignity and indignity have a profound effect on our developing understanding of our value and worth. During childhood, when we are vulnerable and dependent on others for our sense of well-being, we need our caretakers' ongoing love and attention to set the stage for the development of our fledgling dignity.[2] If we experience the opposite—abuse and neglect in its myriad forms—we start our lives doubting our worth. Instead of developing a sense that we are good, valuable, loveable, and worthy, our inner world becomes dominated by a sense of inadequacy, badness, and fears of being defective. This primitive and childlike way of making meaning about ourselves becomes embedded and lasts into adulthood. Unless we do the work necessary to replace the childlike understanding of what happened to us early in our lives with an adult perspective on it, we can remain haunted by self-doubt and continue to feel uncertain of our worth.[3]

The way we treat our children matters. Their brains are vulnerable to abuse and neglect because they are in a constant state of development. Bruce Perry, a specialist in child trauma who trained both as a psychiatrist and as a neuroscientist, has documented his findings of the effects of childhood trauma on brain development and the quality of life of abused and neglected children.[4] He was one of the first researchers to debunk the myth that children are naturally resilient and bounce back no matter what they suffer. When undergoing treatment, traumatized children often used to be medicated for depression or anxiety disorders, but what they had gone through was neglected. Perry has spent his career developing innovative treatment protocols informed by his knowledge of neuroscience and the effects trauma has on normal brain development. Admittedly, a majority of the children he has cared for suffered severe abuse and neglect—some were raped or witnessed the murder of a parent—but he tells us that from their experiences we can learn an enormous amount about the psychological needs of children in general.

Although most of us have not suffered such severe psychological trauma, most of us did experience wounds to our dignity during the psychologically formative years of our childhood. Because of pervasive ignorance about the fragility of humans' emotional worlds, few of us have developed an awareness of the lasting effects of the psychological harm done to us or of the psychological harm that we have done, especially to our own children, who need our care and loving attention.[5]

Children's awareness of their worth begins with the way they are treated early on by their caretakers. If their dignity is violated more than it is honored, they will live in a constant state of doubt about their worth.

Let me be clear. When we are ignorant of the effects that our behavior has on others, and if our culture perpetuates and enables that ignorance, we will unknowingly do harm to one another. Even if we know that we are doing harm, in the absence of explicit societal norms to correct our behavior, we may continue to do harm. Because of this complex interaction of ignorance, denial, and societal taboos against discussing emotional trauma, it is no wonder that we have all experienced some kind of violation in our early lives. And it is no wonder that the caretakers who are responsible for violations of children's dignity are either unaware of the violations or are ignorant of how to nurture it.

What is important here is to know the dignity violations we experienced. To know them is to name them, to give them legitimacy and validation. And knowing them is the first step toward healing. The problem, as Jennifer Freyd points out, is our strong loyalty to our caretakers, especially if they are parents, which makes it difficult to view them as anything but good.[6] Breaking through this loyalty is crucial if we want to get to do the healing necessary for us to recognize and accept the harm that was done to us and that distorts our understanding of our inherent worth.

The purpose of examining our early experiences is not to place blame or to make our caretakers feel bad. The purpose is to uncover the truth about what happened to us or, more the point, to uncover the untruth about our unworthiness.

Having an awareness of our early imprints of indignity enables us to identify vulnerabilities in our adult relationships. The early wounds set the stage for our relationships later in life. If we think of relationships as a source of pain (violations of our dignity) rather than a source of safety and comfort, we will have a hard time with intimacy. We become preoccupied with protecting ourselves from others (the default reaction) rather than connecting with them.

The purpose of identifying our early experiences with indignities is to show us where we may have problems in our relationships later on and, in particular, where we may run the risk of unknowingly violating the dignity of others. Our early violations often create blind spots that enable us to unconsciously justify hurting others.

Add to these default reactions the experiences we have all had growing up in a time when most socializing agents (parents, teachers, religious leaders) are blind to the negative impact they have on children, and it is no wonder that our relationships are a mess. No matter what culture we were raised in, the awareness and practice of what is acceptable treatment of one human being by another is at a fairly primitive level. The combination of our evolutionary (genetic) predispositions and the negative and traumatic experiences that we have all endured leaves us with much to learn about healthy and dignified relationships.

As important as it is to know about childhood violations of our dignity, it is just as important to identify the ways in which our dignity was honored and nurtured throughout our childhood. I have found that people are more aware of the ways their dignity has been violated than the ways it has been recognized, honored, and nurtured. Painful, demeaning treatment apparently makes a more lasting imprint than being treated well does. If we start out feeling unworthy and never have an opportunity to challenge that childlike way of making meaning about who we are and the truth of our worthiness, we can unconsciously carry around that distorted belief throughout our lives. Early emotional wounds to our dignity have tremendous power to keep us locked into a perpetual state of self-doubt, even when there is evidence to the contrary.

I heard recently about a man who was awarded a prestigious employee-recognition award for his contributions to his company; by all accounts, the award was a clear affirmation of his dignity. Yet when asked how he felt about receiving the award, he said that he still felt like a number, that he wasn't really seen or recognized for who he was. Without an internalized belief in his own worthiness, unless his wounds from the early imprints to his dignity were cleansed and healed, he would not be able to appreciate any validation of his worth from the outside, no matter how much recognition he was accorded. The uncared-for and untreated wounds to his dignity demand acknowledgment and attention. Recognizing the need for professional help—from a therapist or counselor of some sort—is the first step on the road to recovery.

After dignity wounds and their effects are tended to, it seems to help to develop an awareness of the forgotten ways our dignity was honored during childhood. Perhaps a special aunt or uncle, teacher, or next-door neighbor, or, for that matter, a much-loved childhood pet, provided validation and recognition. My experience has shown that once people have taken care of the internal wounds, they can even recall times that the perpetrators of violations treated them well. Those experiences were obscured by the overwhelmingly negative emotions that violations to our dignity create.

The negative power of unhealed wounds to our dignity can keep us in a frozen state of self-doubt, preventing us from accessing the positive power that is at our disposal once we see and accept our value and worth. We must tend to our dignity wounds if we want to grow and develop, if we want to abandon a self-protective stance in order to move forward and be open to creating relationships in which we feel safe.

# 4

## Acknowledgment

Give people your full attention by listening, hearing, validating,
and responding to their concerns, feelings, and experiences.

In the early days of my career, I felt a little uneasy at international
conferences because I was often one of a handful of women participat-
ing. Men always led the discussions, and at times it was daunting to
speak up. Sometimes it took a lot of inner strength and self-persuasion
for me to raise my hand.

I attended a small meeting of international experts, from dip-
lomats to academics—maybe thirty people were there—who had
gathered to identify and discuss the cutting-edge issues in the field
of international conflict. A middle-aged man whose organization
was responsible for convening the meeting chaired the sessions. In
the first session on the first day, he asked everyone to think about
the critical issues that needed to be examined. He said that he would
give everyone a chance to contribute, and then we would discuss the
collected comments one by one.

I knew what I wanted to say, and I waited for several other people
to speak before I raised my hand. When the chair acknowledged me,
I said, "I think one of the critical issues we need to look at is the role
dignity plays in international relations. In fact, it's really 'indignity'

that I want to discuss. My experience tells me that the way we treat one another matters, and when people feel their dignity has been violated, they will go to war, if necessary, to regain it."

The chair thanked me and proceeded to gather other comments from the audience. Then he started a discussion about each issue. When it came to my issue of dignity, he looked up from his list and said to the audience, "I think we'll pass on this one," and moved on to the next issue.

I was stunned. I couldn't believe my ears. I felt utterly humiliated.

I sat in my seat, unable to listen to the discussion that followed. Part of me wanted to run away, but I decided that to maintain my dignity, I had to stay seated and focus on the discussion. I had to grapple with this ambivalence for the rest of the two-day conference.

It did help that several people came up to me during a break and said that they were sorry that the chair had passed on my comment. But the damage to my dignity had been done. And the value of my participation in the conference was undermined.

Conflict stays alive when people do not feel acknowledged and when their voices are not heard. Humiliation creates anger and frustration, feeding the conflict, if not escalating it.

The chair could have avoided humiliating me by instead asking me why I thought the issue of dignity was important. He could have given me the benefit of the doubt (another way to honor dignity) by assuming that I had a point that was worth discussing. Instead, he jumped to the conclusion that my idea had no value; he didn't seek to understand why I had raised the issue (another dignity violation). Had we had an exchange about it, we both might have decided that the topic was too big to discuss, which would have saved me from humiliation. Mind you, this episode happened well over a decade ago, when the issue of dignity was not on anyone's mind (except mine).

Another example of the devastating effect of not being acknowledged happened during a dialogue between Israelis and Palestinians that I facilitated with my colleagues. The participants were talking about a new way to approach one of the most intractable issues that divided the two communities. Some forward momentum seemed to be building. Then one person said, "It sounds like a good idea, but we will never be able to sell it to our people." The discussion continued as if the man had never spoken. He intervened a few minutes later, saying the same thing. "It just won't work. It's too far a stretch for them." Someone from the other side finally said, "Let's keep working on it anyway. There are some interesting ideas on the table." But no one asked him to explain why he felt the way he did.

The concerned man pushed his chair away from the table, crossed his arms tightly in front of his chest, and withdrew mentally from the discussion. I watched him for a few minutes, wondering what I should do. Before I could say anything, he got up from the table and left the room. When I went to check on him, I found him sitting on a small couch in the hallway. His breathing was irregular, and I could see that he was in distress.

"I think I'm having a panic attack," he said. "I can't take this discussion anymore. What's happening in the room is the same thing that happens back home. My people are treated as if we are invisible; our concerns are not taken seriously. If I'm feeling this way, you can imagine how my community feels. This lack of respect for who we are and what we have been through is making me sick."

We talked for a few more minutes. I acknowledged his frustration over not being seen or heard, and empathized with the anger that he was experiencing. He calmed down and thanked me for coming to check on him. Although I didn't realize it at the time, my going after him and expressing concern for him was the acknowledgment that he needed to pull himself together. Someone cared. I gave him my full attention.

At the time, I wasn't thinking about the incident as a violation of his dignity, but after many more similar experiences in other dialogues, I began to recognize the pattern—the emotional storms—as the result of not being acknowledged, of being treated as insignificant. These assaults to dignity that underlay political discussions impeded the search for peace.

Extending an acknowledgment to someone who has suffered an indignity is easy enough—and doing so is more powerful than most of us realize. All that is needed is to say simply, "That must have been terrible," or "I'm sorry that you had to go through that," or "I hear what you are saying. Let's discuss your concern." Whatever form the acknowledgment takes, it has a big impact. Acknowledgment doesn't mean that you agree with the person, or that what she says is necessarily the truth. It means that as one human being to another, you see and acknowledge how difficult it is to endure an indignity. Acknowledgment has a deep healing effect on someone whose sense of worth has taken a hit.

5

# Recognition

Validate others for their talents, hard work, thoughtfulness,
and help. Be generous with praise, and show appreciation
and gratitude to others for their contributions and ideas.

A participant in one of my workshops, Brian (not his real name) told a story about approaching his supervisor, Tom (not his real name), with a solution to a problem that had long plagued the company—a problem that Tom had to deal with every day. But Tom didn't have the authority to make the necessary changes; so he asked Brian to develop a PowerPoint presentation to show to colleagues in senior management. Brian was pleased that his boss recognized his insight, and he knew that if the company adopted his recommendation, not only would the problem be fixed, but the solution would save the company a considerable amount of money. Success would look good at Brian's quarterly review and just might give him an advantage at promotion time.

Brian read the current literature on the topic and found significant evidence that his policy recommendation would vastly improve the efficiency of a procedure that had caused a multitude of problems for the company. He worked for two weeks on the project and at last felt ready to show Tom the forty-five-minute presentation. He knew Tom would be able to show it to his colleagues and make a strong case.

Tom thanked him after he saw it, and expressed excitement about the idea and the potential that it had to make a positive change in the company. He said that he would set up a meeting with his colleagues and let Brian know what happened. Brian left Tom's office feeling that he had reason to feel proud.

A week passed, and Brian had not heard anything from Tom. Planning to ask him after lunch, he headed out, only to see two senior managers waiting at the elevator engaged in a lively discussion. He couldn't help but overhear one of them going on and on about what a brilliant idea Tom had introduced and how much it was going to benefit the company. His interlocutor said, "I didn't think Tom had it in him." Brian listened more carefully, and the more he heard, the more he certain he was that they were talking about his idea.

At first he felt sick. Then the sick feeling changed to outrage. How could Tom take credit for his idea? How unfair it was not to get the recognition he deserved! Brian bought a cup of coffee in the lunch room, sat down, and calmed down. He felt sick again, but he was also left with a dilemma. If he confronted Tom, he might lose his job. Positions like his were coveted in the company. Even if he kept his job, Tom, one of the top leaders in the company, could make his life miserable. Brian concluded that he should keep quiet.

Back in his office, he found an e-mail from Tom saying that the meeting had taken place that morning and that the committee had decided to implement the policy recommendation. That was it. There was no mention of Brian's role—no recognition whatsoever.

Brian phoned his wife. He needed a sympathetic ear, if not a reality check. It is bad enough to live with the internal resentment that a violation of dignity creates, but to keep it inside is even worse. Brian needed someone to tell him that what had happened was wrong.

Although this story is a clear example of being treated unfairly and not being given due recognition and acknowledgment for a job well done, the dignity violations did not stop there. This is another case of a cluster violation; many essential elements of dignity are involved.

Brian felt that there was no way to hold Tom accountable for his actions. Because Tom had the power to do even more harm to Brian if Brian fussed about Tom's failure to recognize his contribution, Brian felt resigned to doing nothing—a common response not only in work settings but in any setting where a hierarchical authority structure exists and people do not feel that they can bring up sensitive issues.

Brian's psychological safety was also violated: he did not feel that he could speak up about the issue for fear of retribution. He did not think he had the freedom to confront Tom. And when this kind of domination and control feels like an inevitable consequence of the hierarchical power structure within an organization, resentment accumulates. It would take only a small incident in the future for Brian to completely lose his temper with Tom. Resentment always finds an outlet, whether it is appropriate to the situation or not.

What about Tom? I can't help but wonder why Tom, who was in a significant leadership position in the company, decided to take credit for Brian's idea. Was he not aware of how Brian would feel? Did he think, because of his position of power, that he could claim credit for a subordinate's good work? Whether Tom was ignorant of the negative emotional impact of the insult to Brian's dignity or whether Tom willfully abused his power to serve his own ends, I consider his actions to represent a failure of leadership. What is the role of a leader if not to set an example for those he is responsible for?

❋

My work has brought me close to leaders of all kinds—in political, business, educational, military, and religious positions. Many of them have described a common leadership challenge: knowing what to do when emotional issues arise. In spite of the many years spent collecting degrees and intellectual capital, they rarely feel confident when faced with people who are experiencing outrage, who feel that they are being treated unfairly, whose unacknowledged grievances have changed them into fighting men and women. In other words, the leaders also share an area of ignorance: they don't know what to do when faced with people who have experienced violations of their dignity, which are, by definition, highly charged emotional events.

While I was becoming aware of the powerful impact of a dignity violation, I saw how ill equipped most of the people I worked with were to handle an emotional upheaval. Their default reaction was to use their authority and the power of their position to control the overall situation, often leaving the aggrieved people angrier, more resentful, and less willing to extend themselves in their jobs or their roles within an organization.

Another reason why the default reaction for some leaders is to exert authority and control over a volatile emotional situation is this: they are afraid. They have described their fear to me in two ways: they are afraid of being exposed and embarrassed by having made a bad move or having established a flawed policy, and they are afraid because they don't know what to do when someone expresses raw emotions.

I have seen otherwise brilliant leaders get caught in all of the predictable traps associated with not knowing how to handle emotions. They are not bad people who deliberately try to make life difficult for those whom they lead; they simply don't have the knowledge, awareness, and skills to navigate through the emotional turmoil caused by dignity violations, either their own or others'.

The need has never been more urgent for people in leadership positions to be educated in all matters related to dignity—both the human vulnerability to being violated and the remarkable effect on people when they feel that they are seen, heard, understood, and acknowledged as worthy. Treating others well and recognizing their humanity (both their worth and their vulnerability) have incalculable benefits everywhere that human beings cluster: in families, communities, the workplace, and nations. Honoring people's dignity is the easiest and fastest way to bring out the best in them. The opposite is equally true. Treating people as if they don't matter creates destructive emotional upheavals.

What would it look like if leaders were required to lead with dignity? Leading with dignity means that leaders are aware of the emotional volatility of experiencing an assault to one's worth. Leading with dignity would require them to demonstrate what it looks like to treat others as if they matter; it would require them to know how to treat people who have been violated and what steps to take when they themselves have violated others. Leading with dignity also means knowing how to maintain one's own dignity, especially when faced with the many temptations that lure us to do otherwise. People everywhere have told me how desperate they are to see their leaders acting with dignity.

Paul Woodruff, author of *Reverence: Renewing a Forgotten Virtue*, addresses this point. He makes the impassioned plea that we resurrect the notion of reverence—the feeling of being in awe of and aware of a force or forces greater than ourselves.[1]

The experience of reverence is crucial because it provides us with an internal check on our oh-so-human tendency to think of ourselves as superior beings, an attitude that can justify all kinds of harmful behaviors. Without the feeling that there is something greater than ourselves, we risk losing sight of our human limitations. The something greater could be God for those who are religious, or ideals like truth and justice, which captivated the Enlightenment philosophers, or the magnificence of the universe, which is enough to bring Richard Dawkins, an atheist, to his knees.[2]

Woodruff argues that reverence is especially important for our leaders to feel. Basing his insights on ancient Greek philosophy and the teachings of Confucius, he points out that it was important in both Greece and China that rulers exhibit reverence, because otherwise they would be tempted to exert too much control over the people they led. As Woodruff says, "Reverence is the virtue that keeps human beings from trying to act like gods. To forget that you are only human, to think you can act like a god—this is the opposite of reverence. Ancient Greeks thought that tyranny was the height of irreverence, and they gave the famous name of hubris to the crimes of tyrants. An irreverent soul is arrogant and shameless, unable to feel awe in the face of things higher than itself. As a result, an irreverent soul is unable to feel respect for people it sees as lower than itself—ordinary people, prisoners, and children."[3] In ancient China, a ruler whose reign was marred by war, crop failure, or other catastrophes, and whose subjects were starving, miserable, and suffering injustices, had lost the "mandate of Heaven"; he was no longer a model of virtue and practically invited being overthrown.

Woodruff makes another important point about reverence: True reverence is a felt experience, not a belief. People who embody the virtue of reverence are people who feel like doing the right thing; a sense of morality is at the core of what prompts them to behave well.

❀

One of the hallmarks of good leadership, then, is the capacity to feel awe and wonder at something greater than oneself, and this feeling acts as a check on arrogance. Being able to feel reverence has a further effect: it creates the sense of humility necessary to avoid the temptation to abuse one's power by harming, exploiting, and disempowering others.

Looking at leadership in this way, we can see that many who hold positions of power are not leaders in a true sense. They do not exemplify the virtues of wanting to do right and accepting their (and others') human limitations. Abuses of power are rampant in political, business, educational, and faith communities. There are enough recent scandals to prove the point, from pedophilia among the priests in the Catholic church, to myriad unethical schemes on Wall Street, to countless acts of tyranny in political regimes all over the world. From Woodruff's point of view, leadership is in a state of crisis.

Paul Woodruff is right: we are in desperate need of leaders who feel reverence. Reverence is a necessary condition for leading with dignity because the feeling of awe that reverence brings also gives us the capacity to recognize the value of others, as well as react with shame when we fall short and harm them.

The concept of reverence casts a whole new light on the work of leaders. I have made the case that a leader should embody a level of emotional sophistication in order to fulfill the obligations of her role. Reverence is the starting point, and understanding the complex role that dignity plays in emotional vulnerability and well-being is closely related to it. Without the feeling of reverence, it would be difficult to understand the emotional force behind the power of dignity.

My point is simple: If leaders want people to buy into and support their ideas about governance, understanding how to keep them feeling emotionally well balanced is fundamental. Treating them with

THE TEN ESSENTIAL ELEMENTS OF DIGNITY

dignity creates not only inner stability but a sense of belonging to something bigger than themselves, whether it is an organization, a corporation, or a nation. Those who feel that they are seen, included, and recognized for their contributions are more willing to extend themselves for this greater good than are people who feel dismissed. That is why Woodruff believes that the feeling of reverence holds communities together. If a leader has both the feeling of reverence and a knowledge of dignity, that leader becomes a moral authority, able to guide the people in a particular group as they all struggle to become better human beings.

# 6

## Fairness

*Treat people justly, with equality, and in an evenhanded way according to agreed-on laws and rules. People feel that you have honored their dignity when you treat them without discrimination or injustice.*

At the end of a two-week trip to Sri Lanka, my colleagues, William Weisberg and Sydney Silva, and I had seen more than enough human tragedy. The Tamil Tigers and the government of Sri Lanka were at war. We went from one part of the country to the other, interviewing potential participants for a workshop that would bring together representatives of the Tamil, Muslim, and Sinhalese communities for dialogue. Just getting around Sri Lanka, with the incessant traffic jams, pollution, and intolerable heat, left us exhausted by the end of every day.

One day, when we were stopped at a light, I noticed a truck full of Sinhalese boys in green military uniforms. Like any other boys, they were playfully pushing and shoving one another. I turned to Sydney and William and said, "Do you think they have any idea what they're in for? Most likely, in a day or two they will be on the front lines in the north fighting the Tamil Tigers. "

Sydney replied, "It's not fair, is it? These boys deserve to be playing cricket, not fighting a war."

We sat in silence, watching the truck as it pulled away.

On one of the last days of our stay, tired and emotionally overwhelmed by everything we had seen firsthand, we met with the director of Catholic Relief Services (CRS), a nongovernmental organization in the capital city of Colombo, to see if he could help us gain access to the Tamil Tigers in the north of the island.

At the time, CRS focused many of its relief efforts on children, and we were in a room whose walls were covered with posters of children in refugee camps, in small villages, and in makeshift schools in a war zone. While listening to Sydney explain our dilemma to the director, I examined the faces of the children in the posters. My eyes stopped at one little boy—he must have been about five—standing in a line reaching for a plate of food. I could feel something inside me snap. All the sadness I had kept in check over the past two weeks overtook me, and I started to cry.

My need to leave the room took me into the hallway. I stood there, looking left, then right, not knowing where to go. Turning to the entrance door, I came face-to-face with an elderly nun. She took me by the hand and led me into her office, where she sat quietly, her hands folded in her lap, while I wept. After a few minutes, I looked up at her and said, "It's the children."

She took me in her arms and held me, and I let myself be comforted. When I stopped crying, I could only say, "Thank you."

We both stood up. She took my face in her hands, looked at me, and smiled. I stepped away, walked toward the door, and then stopped. Looking back over my shoulder—in a flash of a second—I caught her wiping away tears with the back of her hand. With a smile on her face she waved at me, and I left.

With the clarity that comes after a good cry, I knew then that children suffer the worst of human injustice. Whether they are treated unfairly by the circumstances of war or by wars within families, it is wrong for children to suffer. It is the ultimate indignity to cause them harm, especially if we have it within our power to prevent it. No child deserves to suffer indignities that have the potential to distort their sense of safety and worthiness for the rest of their lives.

The story of my visit to Sri Lanka focuses on the injustices perpetrated against children, but the issue of fairness underlies just about any violation involving any of the other nine essential elements of dignity. Is it fair, for example, when someone is excluded, unrecognized, misunderstood, or treated as inferior? Is racism fair? Or sexism, inequality, religious intolerance, homophobia, classism, or the indignities forced on people with disabilities? Is it fair to be on the receiving end of behaviors that stem from ignorance? Is it fair that the issue of dignity has for so long not been addressed? Is it fair that our reluctance to shed bright light on our emotional vulnerabilities has cost us all so much internal pain and suffering? Is it fair that leadership has been so corrupted—that power in the hands of the wrong people can destroy so many lives? Is it fair that we have lost sight of other forms of power—the expansive power that is unleashed when we recognize the dignity of others?

Our reaction to being treated unfairly runs deep and affects us early. Children seem to recognize unfairness instinctively. Imagine a scenario in which a mother is giving her two young children candy. The first thing that most children do is examine how much candy the sibling received in comparison. Given even the slightest chance of an imbalance, a child will point it out with a fair amount of indignation.

Fairness is an issue all through life, of course, and applies at a much larger scale than the distribution of candy. Is it fair, for example, that the economic, social, cultural, and ecological integrity of coastal communities in Louisiana, Alabama, Mississippi, and Florida were endangered by a massive oil spill in the Gulf of Mexico? People whose livelihoods are dependent on fishing or other ocean-related industries are outraged by what has happened. Is it fair that millions of innocent people will suffer indefinitely from this disaster? What about the marine life and other wildlife that have been affected? What about the short-, medium-, and long-term effects on the environment?

Although most of the attention in this book has been focused on human indignities, the idea of dignity doesn't stop with human beings. The indignities that we perpetrate on other species and on the environment deserve just as much attention. Thinking about how we treat all aspects of life on the planet, and the planet itself, in terms of dignity is a way to connect ourselves emotionally to the world around us. By allowing ourselves to feel the consequences of our actions—of the harm we do not only to each other but to all life—we will gain a different view of climate change and other human actions that affect the planet. Our feelings about fairness will expand to include the natural world, and we won't need to be reminded that all living things, not just humanity, are invaluable, priceless, and irreplaceable.

# Benefit of the Doubt

Treat people as trustworthy. Start with the premise that
others have good motives and are acting with integrity.

One of the greatest demonstrations of human dignity I have come
across in the many years that I have been grappling with the subject
was when Nelson Mandela walked out of prison after twenty-seven
and a half years of confinement by the apartheid government in South
Africa and announced that he had no anger in his heart toward
whites. That, as I said earlier, earned him enormous respect. In his
autobiography, *Long Walk to Freedom,* he describes how he felt at his
first press conference, a day after his release on February 12, 1990. In
response to a question, Mandela said, "I knew that people expected
me to harbor anger toward whites. But I had none. In prison, my
anger toward whites decreased, but my hatred for the system grew.
I wanted South Africa to see that I loved even my enemies while I
hated the system that turned us against one another." He declared
that there was a middle ground between white fears and black hopes,
and the African National Congress would find it. "Whites are fellow
South Africans," he said, "and we want them to feel safe and to know
that we appreciate the contribution that they have made toward the
development of this country."[1]

Yes, this was one of the greatest acts of humanity of my lifetime, and I am not sure that the rest of us will ever completely understand how he was able to let go of the very human reaction of anger, if not revenge, toward those who had treated him so badly. What I keep coming back to is dignity—his understanding of his *own* dignity. Here is what he said: "Prison and the authorities conspire to rob each man of his dignity. In and of itself, that assured that I would survive, for any man or institution that tries to rob me of my dignity will lose because I will not part with it at any price or under any pressure."[2]

Because Mandela's sense of his own value and worth was so deeply ingrained, he never lost sight of the inherent value and worth of others, no matter how badly he was treated. His sense of dignity is the source of his humility. It is humanity itself that he respects—not just his own but that of every human being. Keeping that truth to the fore is an astonishing human achievement.

Near the end of his book, Mandela writes that he always knew that every human heart, deep down, held mercy and generosity. "No one is born hating another person because of the color of his skin, or his background, or his religion. People must learn to hate, and if they can learn to hate, they can be taught to love, for love comes more naturally to the human heart than its opposite. Even in the grimmest times in prison, when my comrades and I were pushed to our limits, I would see a glimmer of humanity in one of the guards, perhaps just for a second, but it was enough to reassure me and keep me going. Man's goodness is a flame that can be hidden but never extinguished."[3]

Mandela's respect for humanity never faltered. It seems to have been the one anchoring truth that let him give even the most brutal of his captors the benefit of the doubt. Under circumstances that would drive most of us to demonize those who do us harm, Mandela didn't. He never doubted his captors' capacity for goodness and tenderness.

It is as if he recognized something in them that they were not able to see in themselves. He saw something bigger, something greater, something sacred—he saw their dignity. What is the lesson here? Is he showing us that in order to give others the benefit of the doubt, we must know that we ourselves are worthy of dignity? Is it our own self-doubt, our own feelings of unworthiness, that fuel our cruelty toward others?

Although I have admired Nelson Mandela only from afar, I have met the other moral giant responsible for the nonviolent transition of South Africa to a nonracist democracy: Archbishop Emeritus Desmond Tutu. Archbishop Tutu and I worked together in fall 2005 on a BBC project aimed at addressing the unhealed wounds of the conflict in Northern Ireland. Even though the Good Friday Agreement that brought an official end to the conflict between the Catholics and Protestants had been signed in 1998, seven years later tensions still could be felt in the everyday lives of people, especially those who had lost loved ones in the conflict. The BBC recognized the need to deal with the losses to help put the past to rest. What the BBC proposed was a three-part television series, *Facing the Truth,* which brought victims and perpetrators of the conflict together in dialogues. Who better than Archbishop Tutu to preside over the encounters? He was the man who had chaired South Africa's Truth and Reconciliation Commission, which helped thirty thousand of his countrymen and countrywomen tell their stories of loss and suffering.

Later in this book I describe those days together in the tiny village of Ballywalter, forty miles outside Belfast, in a beautiful home owned by the Dunleath family. What I want to underscore here are the similarities between Nelson Mandela and Desmond Tutu in their capacity not only to recognize and accept the inherent dignity of all human beings but to embody it. For them, giving people the benefit of the doubt—no matter how heinous their behavior—appears to come naturally.

The BBC asked me to co-facilitate the encounters with the archbishop, along with Lesley Bilinda, the widow of a victim of the genocide in Rwanda. Even though I had spent many years working with parties in conflict all over the world, this project terrified me. The focus of the dialogues that I had facilitated in the past was political in nature. Although I was often frustrated with those encounters because I felt a need to go deeper into the emotional maelstrom that conflict creates for human beings, I never dreamed that my wish would come true in this real yet extraordinary way. Now I would be facilitating a first-time encounter between a woman who had lost her husband and the man who had killed him. They would be sitting across the table from each other, just an arm's length apart.

I was nervous about two things. Would I be able to facilitate such an emotionally volatile encounter? And would I be able to open my heart to that man and others who were responsible for the sometimes brutal killings? Would I be capable of giving them the benefit of the doubt?

The BBC production crew met at the Dunleath house a few days before the programs were filmed. Apart from needing time to get over jet lag, those of us on the facilitation team needed to get to know one another and to create a common understanding of how to go about the enormous challenge ahead.

Looking back on those prefilming sessions, I am sure that we were all feeling the safety that comes with the archbishop's unconditional acceptance of us—his unspoken honoring of our dignity by giving us all the benefit of the doubt. When he sets those eyes of his on you, and you feel his undivided attention, the experience is overwhelming. Lesley Bilinda talked about her despair when her husband died, and she described her driving desire to go back to Rwanda to find the men who killed him. I especially remember surprising myself by talking openly about my inability to have children, a loss I rarely discuss. Why it came up then, I really don't know. Perhaps it had something to do with being seen, heard, listened to. Openness is one of the many wonderful consequences of giving people the benefit of the doubt: they feel safe to be vulnerable.

Although unconditionally giving people the benefit of the doubt is a quality that both Nelson Mandela and Desmond Tutu share, they have something else in common: they accept their own worthiness. Although the prison guards at Robben Island tried to strip the prisoners of their dignity, Mandela knew that no one could rob him of his, which helped him survive the ordeal. In the movie *Invictus*, which portrays Mandela and the way he won the hearts of the white Afrikaners by supporting and strengthening their rugby team, it was pointed out that his conviction about his dignity was inspired by a nineteenth-century British poet, William Ernest Henley, who wrote the poem "Invictus." Here are the words that helped Mandela maintain his dignity throughout those twenty-seven-plus years:

### Invictus

> Out of the night that covers me,
> Black as the Pit from pole to pole,
> I thank whatever gods may be
> For my unconquerable soul.
>
> In the fell clutch of circumstance
> I have not winced nor cried aloud.
> Under the bludgeonings of chance
> My head is bloody, but unbowed.

Beyond this place of wrath and tears
Looms but the Horror of the shade,
And yet the menace of the years
Finds, and shall find, me unafraid.

It matters not how strait the gate,
How charged with punishments the scroll.
I am the master of my fate:
I am the captain of my soul.

Archbishop Tutu has a different way to anchor his dignity, but the result is the same. He feels worthy no matter what happens to him, and he, too, is able to give others the benefit of the doubt.

At the end of a particularly hard day of filming by the BBC, the archbishop wrapped up the encounter by telling the two participants that he was "deeply humbled" by them, that their ability to reach out to each other instead of throwing missiles reinforced his belief in the goodness of humanity. His closing words to them were "Thank you for being vulnerable." The two participants beamed with pride. The archbishop was right—they were at their best that day.

I pondered his ability to preside over such a difficult session and to wrap it up so that the two participants walked out feeling enlarged in spirit. I asked, "Archbishop, how do you do that? It was such a powerful ending. You obviously hadn't rehearsed it, so where does it come from?"

He giggled and pointed to a light in the ceiling.

"What?" I asked.

"I'm like a light bulb. When I'm plugged into the source, I shine."

# Understanding

Believe that what others think matters. Give them the chance
to explain and express their points of view. Actively listen
in order to understand them.

"Once we get straight the equality between men and women, many of the other 'isms' will disappear."[1] I hope Shulamuth Koenig, director of the People's Movement for Human Rights Learning, is right in saying so, but at this point in human history, we still have a way to go before we find out. Koenig believes that it is the quality of the relationship between men and women that sets the stage for all other relationships. If children see their parents treating each other with dignity, then they will develop an imprint in their brain about how all others should be treated. The idea is worth pondering.

Unfortunately, it is all too common today to see gross inequities in the treatment of men and women, especially in the workplace. I have consulted for an organization whose leadership hierarchy has been historically dominated by men. The organization has made a concerted effort to include more women in high-ranking roles. The workshop I conducted there was designed to use the dignity lens to identify dysfunctions that were making it an unpleasant place to work.

In the beginning, I explained to the participants that if they wanted an easy way to see both dysfunctions and leadership failures, all they had to do was look for where people were suffering; there they would probably find a stockpile of dignity violations. The first issue brought up was the suffering of female employees. The women pointed out that not being given leadership positions just because they were women was a violation of their dignity. Among them, the women accounted for a fair number of unprocessed stories and a great deal of unacknowledged suffering. They felt that whenever they brought up a grievance, the men rushed to fix the problem without giving them a process through which they could speak of their indignities and be heard. What they were seeking was recognition and understanding, and the men, they said, had a hard time listening.

My sense was that the men in the group listened to the stories the women had to tell, but they didn't know what to do about them. They were in a difficult position. They felt uncomfortable hearing about the raw experiences that so many of their fellow employees had endured. The stories had a lot of pain attached to them, and the men weren't sure how to address that pain. It was easier for them to rush into the problem-solving mode than to sit and feel.

One incident during our workshop indicated the lack of understanding that prevailed because these men and women did not know how to have a conversation without someone getting hurt. Even though the pain and suffering of the women in the company had never before been openly discussed, the issue came up constantly in oblique ways. And because the women had had to stifle so much resentment about being treated as second-class, the tone of voice in most discussions was negative. The negativity was never directed at anyone in particular, but it was in the air. It seemed to put the men on the defensive, even before an issue was put on the table for discussion. The men were often waiting for an attack—they feared their dignity was on the line.

The sad thing was that the men, at least the ones in the workshop group, were concerned about the plight of the women. They, too, felt that it was unfair that women were not in positions of power. But because of the volatility of the issue and the emotional charge that it carried, everyone at the table was on the verge of becoming a dignity violator. One false move, and both the men and the women were prepared to strike.

When one woman spoke with that edge to her voice about how difficult it was for her to bring up issues of power with her boss, one of the men in the group jumped on her. "I know who you're talking about, and you've got him all wrong. He definitely would want to hear your concerns. You are misrepresenting him, and you're only adding fuel to this fire."

Chin jutting forward, she leaned into the table and said, "Are you trying to talk me out of my experience? Do you really think you know better?"

This was a perfect moment to demonstrate the dignity model. Seeing the situation through the lens of dignity could avert a discussion full of reciprocal violations.

I asked the arguers if they could stop for a second and if they would be willing to use the tools of the model to have a different kind of conversation about the issue. They both agreed. The other participants started whispering to one another, obviously fearful and anxious about what I was about to do. I felt nervous myself, but I knew that a different kind of discussion might shift the dynamics and enable the two to speak to each other in a way that would allow them both to feel that they were being seen and heard.

I asked Mike (not his real name) if he would be willing to rewind the tape of what had just happened. I suggested that instead of acting defensive and accusing her of misrepresenting her boss, he tell Monica (not her real name) that he didn't understand what she was saying. I encouraged him to ask her if she could say more about what it was like for her to interact with her boss. So he did. He said, "Monica, I really didn't understand what you were saying. My experience of your boss is completely different. Would you please tell me more about what it's like for you to approach him? I really want to understand your experience with him."

Monica sat quietly for a minute, collecting herself, and then told a story about how she, a woman and a person of color, faced the power structures of the organization. She explained, in the most vulnerable way, what it was like growing up in a world where she was smart—probably smarter than most—yet the shame that was attached to her identity kept her from standing up for herself. Her voice was not heard. She said she had been fighting shame all her life, that the anger and the resentment felt like poison inside her, yet she didn't know what to do with it. More than anything, she wanted to find the internal strength—the dignity—to believe in herself and to act accordingly.

"I'm sick to death of being angry," she said. "Mike, I want you to understand me, and I want to understand you. And I want to tell you about me in a way that you can hear, so my story will pique your curiosity, if not your compassion, about me and what my life is like. I want you to see me as a human being with the same dignity that's in you." With tears streaming down her face, she said, "Give me the benefit of the doubt that what I tell you is true for me, and I will have no choice but to do the same for you."

The man got up from his chair, went over to where she was seated, and embraced her. The healing conversation had begun.

It is tempting to rush to judgment—to think you know better—when someone says something that doesn't jibe with your understanding of reality. Often we go on the defensive instead of stopping ourselves and considering that maybe there is more to the story than the information at hand. The simple act of saying to a person, "Please say more about your experience," instead of rushing to "You're wrong," completely shifts the dynamics of a conversation. The person, instead of being put on the defensive, can now feel seen and heard. Nothing is lost by asking for more information. Whatever the response, it will enrich your understanding, and the person will feel more understood because you asked, because you showed that his or her perspective matters to you. Seeking understanding is one of the easiest ways of honoring dignity. Allowing people to feel understood is just one short sentence away: "Tell me more."

# 9

## Independence

*Encourage people to act on their own behalf so that they feel
in control of their lives and experience a sense of hope and possibility.*

War is the inevitable consequence of choosing to resolve differences by force. Domination and control are the goals of war. Prisoners may be taken, and we all know that torture is not uncommon when they are questioned, even though international law forbids it.

At the end of a conflict-resolution workshop between warring parties held at Harvard, I had an opportunity to speak privately with one of the participants, someone who played a leadership role in his community, about the time he had spent in prison as a political prisoner. I wasn't planning to have an intimate conversation. We were seated next to one another at dinner and made small talk for quite a while. He asked me questions in an effort to get to know me, which doesn't happen often with influential people—they are usually busy telling you all about themselves. At one point, he jokingly mentioned his time in prison.

I was surprised to hear that he had been a prisoner and asked him what it was like. For a few seconds he studied my face. I am not sure what he was looking for, but after a short pause, he started to tell me stories about the day-to-day abuse by guards who were younger than his children and about being humiliated during interrogations.

One thing he told me will stay with me forever. I asked him how he reacted when he was being humiliated. He said: "From the time I was a little boy, my mother told me that I should never react when someone treats me badly. I should not lose my temper. She said it didn't matter whether someone hurt me with words or with a stick. Never let them know they are hurting you. And that wasn't all. She said I should always maintain a smile on my face because you never want to give the person the satisfaction of thinking he has injured you. *You are the one in control of your dignity, not the person hurting you.*"

The last sentence struck home. I added the emphasis mentally. I realized that what he said was true: ultimately we are the ones in control of our own dignity, even if we are being held against our will. My dinner partner's words reminded me of a quotation of Eleanor Roosevelt's: "No one can make you feel inferior without your permission."

At the end of our conversation, after listening to him intently for nearly an hour, I told him, "I don't know how you did it. I don't think I would have had the strength to endure what you did with such dignity."

He looked at me and began to cry. It was only for a second, but he cried. He quickly composed himself, momentarily put his hand over mine, and then turned to his aide, sitting on the other side of him, who was tapping him on the shoulder. "Mr. X——, the waiter wants to know what you would like for dessert: crème brûlée or chocolate gateau?"

This story is about the inhumanity of war and how it strips us of our freedom, but the impulse to control and dominate others is not confined to the battleground. The imperative of war—to use power to deprive others of power—is all too often the imperative in our everyday lives as well.

Minority groups often find their freedom restricted by the imposition of the norms of the dominant culture, and that restriction has been enough to incite serious political unrest. Members of the minority group want to express themselves and their unique identity freely. Without coercion, it is difficult to repress a group's desire to speak their own language and practice their own religion—to choose how to live their lives in a way that gives them meaning.

We can see the impulse to restrict the freedom of others in most of our relationships, not just at the national level. Imagine the domineering husband who wants to control his wife, the overprotective mother who fails to see the importance of letting her daughter experience herself as an independent being with her own dreams to fulfill, or the executive who refuses to hear the concerns of employees out of fear of losing authority.

The real problem with domination is that it takes away a person's independence, thereby violating one of the essential elements of dignity. Like any other dignity violation, taking away a person's independence creates resentment. Resentment contaminates relationships, setting the stage for the transformation of the other into an enemy. In contrast, using power to empower others not only guarantees them their freedom and protects our own but is a step toward fulfilling our primal longing to be connected with others.

# 10

## Accountability

Take responsibility for your actions. If you have violated
the dignity of another person, apologize. Make a commitment
to change your hurtful behaviors.

A couple of months after 9/11, my colleague William Weisberg
and I were invited to facilitate an interfaith dialogue in New York
City involving several members of the Christian community and
representatives of the Buddhist, Hindu, and Muslim faith traditions.
Tensions were high, and misunderstandings abounded. Everyone was
nervous about how the day was going to go.

To the great relief of the facilitators, most of the morning session,
designed to clarify misconceptions about the various religions, went
well. Not until the end of the discussion did one of the Christian
ministers (male) and a Muslim religious counselor at a major university (female) start a dispute about an interpretation of Islamic law.

The two were sitting at opposite ends of a large oval-shaped table.
They argued back and forth, dominating the discussion, until the Muslim woman leaned into the table, looked the Christian minister in the
eye, and almost in a whisper said, "Are you telling me that you know
more about Islam than I do? Are you telling me that I should go against
all that I have learned and accept your interpretation of the law?"

The minister, taken aback, spoke to her in a condescending tone. "So you think you're the only one who knows about Islamic law? I happen to know quite a bit about it."

Losing her patience, the woman pushed her chair back from the table, crossed her arms, and said, "It's hopeless. I cannot have this kind of discussion if I am not respected for who I am and what I know."

The participants at the table shifted in their seats. My colleague and I looked at each other and called for a break.

While everyone was nervously getting coffee, I saw the Christian minister pacing in the back of the room, head lowered, his hands in his pockets.

After five minutes, we returned to the table. Before anyone else spoke, the minister addressed us all, asking if he could speak first. He sat quietly for a moment, then moved up to the edge of his chair and placed his folded hands on the table in front of him. He looked at the Muslim woman and said, "Mrs. X———, I want to apologize for the way I behaved before we took a break. Of course you know more about Islam than I do. And just to set matters straight, *I* should be the one who learns from *you* about Islamic law. There's something else you need to know about me before you can understand what happened before the break. I am a recovering chauvinist pig, and I have to work at taming my male superior attitude every day of my life. So, please—I ask for your forgiveness."

Mrs. X——— was stunned. She appeared to have readied herself for another attack, and his apology took her by surprise. A few seconds passed before his words registered. She took a deep breath and let out a sigh. Her face softened, and she calmly said, "I accept your apology."

This example is as vivid in my mind today as it was the day it happened. When I saw the minister pacing during the break, I had no idea what to expect from him. My colleague and I were prepared to raise the issue, but not without a lot of trepidation. That he brought his best self forward and took responsibility for what he had done, making himself vulnerable, surprised us all.

We were surprised because many years of convening parties in conflict have shown that apologies do not happen very often. The need to save face, which I will discuss later, is one of the most primal aspects of our evolutionary legacy, and it has caused much harm for us as human beings, holding us back from doing what a part of us knows, deep inside, is right. It takes great strength to fight the impulse to save face. Very few people want to admit that they have done something wrong. The fear of looking bad in the eyes of others and the fear of losing dignity are nearly insurmountable. A paradox is worth noting: Not only did the minister not lose dignity when he admitted that he was wrong and asked for forgiveness, he gained it. He went from looking and sounding arrogant—putting us all off—to becoming noble in our eyes. His display of vulnerability made us all open to him.

I am not sure how neuroscientists would describe the change in our perception. Was it his heartfelt admission of his mistake that made us feel his vulnerability? Was it our mirror neurons that enabled us to open to him, to feel with him? Can our neurons sense when someone's remorse is genuine? Is that why the Muslim woman, whose dignity had been violated just minutes before, was able to accept his apology?

The Christian minister's remarkable intervention recovered his dignity and honored the Muslim woman's at the same time. It was also clear that he had been struggling with his sexist beliefs for some time. There is something powerful about seeing others defenselessly coming to terms with themselves. The minister even went a step further by setting aside his Me, bringing his I forward, and doing what was dignified and right.

Why does his act move us? Maybe it is because we know how hard it is to backpedal, to apologize, to accept responsibility for error. We know that being that vulnerable takes raw courage, that running the risk of looking bad or—what may be worse—feeling bad takes strength and will. Or maybe we are moved because seeing people do the right thing, holding themselves to account for their actions, especially when it is difficult, reminds us of how we want to be. By being moved we are also responding to our best survival instinct—our impulse to comfort someone who has taken a shaky step toward becoming what he or she is capable of being.

PART TWO

# THE TEN TEMPTATIONS
# TO VIOLATE DIGNITY

**Taking the Bait**  Don't take the bait. Don't let the bad behavior of others determine your own. Restraint is the better part of dignity. Don't justify getting even. Do *not* do unto others as they do unto you if it will cause harm.

**Saving Face**  Don't succumb to the temptation to save face. Don't lie, cover up, or deceive yourself. Tell the truth about what you have done.

**Shirking Responsibility**  Don't shirk responsibility when you have violated the dignity of others. Admit it when you make a mistake, and apologize if you hurt someone.

**Seeking False Dignity**  Beware of the desire for external recognition in the form of approval and praise. If we depend on others alone for validation of our worth, we are seeking false dignity. Authentic dignity resides within us. Don't be lured by false dignity.

**Seeking False Security**  Don't let your need for connection compromise your dignity. If we remain in a relationship in which our dignity is routinely violated, our desire for connection has outweighed our need to maintain our own dignity. Resist the temptation to settle for false security.

**Avoiding Conflict**  Stand up for yourself. Don't avoid confrontation when your dignity is violated. Take action. A violation is a signal that something in a relationship needs to change.

**Being the Victim**  Don't assume that you are the innocent victim in a troubled relationship. Open yourself to the idea that you might be contributing to the problem. We need to look at ourselves as others see us.

**Resisting Feedback**  Don't resist feedback from others. We often don't know what we don't know. We all have blind spots; we all unconsciously behave in undignified ways. We need to overcome our self-protective instincts and accept constructive criticism. Feedback gives us an opportunity to grow.

**Blaming and Shaming Others to Deflect Your Own Guilt**  Don't blame and shame others to deflect your own guilt. Control the urge to defend yourself by making others look bad.

**Engaging in False Intimacy and Demeaning Gossip**  Beware of the tendency to connect by gossiping about others in a demeaning way. Being critical and judgmental about others when they are not present is harmful and undignified. If you want to create intimacy with another, speak the truth about yourself, about what is happening in your inner world, and invite the other person to do the same.

When I created the dignity model, I developed the ten essential elements of dignity that we looked at in part I. The purpose of the model was to give a clear picture of what it looked like to honor dignity, on the one hand, and what it looked like to have one's dignity violated or to violate it in others, on the other hand. Describing dignity violations took the concept of dignity out of the realm of abstraction and made it part of our everyday lives. The ten essential elements also provided a language with which to talk about dignity. Knowing the elements enabled people to put a name to what happens in an interaction with someone when they walk away feeling upset, and it legitimized the experience.

After giving several workshops on the ten essential elements, I realized that a piece of the puzzle was missing. I needed to expand the model to include ways in which our hardwired instincts have set us up to violate our own dignity.

The ten temptations to violate our dignity emerged from insights I gained from evolutionary psychology. As I have pointed out, many of our self-protective instincts served our ancestors well when physical survival was the foremost concern. The instinct to fight or flee when someone does us harm is one of those instincts. But there are many more ways in which we are hardwired to react without thinking. The desire to save face, to preserve others' good opinion of us, is a psychological mechanism that evolved to aid survival.[1] Like other parts of our legacy, reacting to save face can get us into more trouble than it saves us from. Covering up when we do something wrong doesn't serve us well in the twenty-first century. This second part of the book will introduce ten ways in which our evolutionary legacy creates problems for us in addition to our impulse to fight and flee. Outdated reactions to present-day realities can set us up to violate our own dignity—unless we learn to control or counteract those reactions.

Evolutionary psychology also tells us that having predispositions doesn't doom us forever. As Jerome Barkow, author of *Missing the*

*Revolution: Darwinism for Social Scientists,* says, "Biology is not destiny unless we ignore it."[2] The first step is to recognize what we are dealing with. Then we can take steps to work around it. Until we learn how our evolutionary legacy influences our behavior, we will be tied to it, arrested in the process of becoming what we are capable of being.

There is so much to learn about our legacy—aspects of our humanity that were instilled hundreds of millennia before we took our first breath.[3] Without this knowledge, we are likely to create significant and destructive distortions in our thinking about who we are and what is possible for us in our lifetimes. Because we are still wrestling with questions of nature versus nurture, and all the related questions about what makes us human, we end up personalizing some things too much and other things too little. In determining what is our fault and what is our responsibility, for example, we often take the path of least resistance, letting the part of us with our primal survival instincts be in charge of our decision making instead of the part of us that is uniquely human, the part of us that knows that the path of least resistance can often be the path to the most destruction.

When someone violates our dignity we tend to personalize it. We feel bad because it happened to us, and we feel bad about ourselves because of it. The violation triggers self-doubt; our inherent human vulnerability makes us respond to being treated as inferior or stupid by denigrating our own worth. But if we understand the inherited mechanisms by which we react to such violations—if we know that shaming probably evolved as a weapon for lowering status, for example—then we will not personalize them so much. In the early development of humans, gaining status was one way to ensure better chances of survival and better reproductive partners—and of course to some extent it still is.[4] Is it any wonder that a powerful internal mechanism evolved to annihilate the competition?

If we become familiar with the instantaneous responses to which we are all predisposed, we will be able to put them in perspective. Through no fault of our own, a great deal of our evolved biology does not work well for us now. The good news is that we can recognize our inherited responses and work around them. We can face the truth about who we are: that we are all born worthy and will die worthy, no matter what happens or what we do in between. The process of becoming what we are capable of being requires that we know and accept our worth.

The ten temptations to violate dignity are all tendencies that we have to manage within ourselves. Sometimes we have to restrain ourselves—hold back from acting on our instincts alone—and sometimes we have to fight the impulse to do nothing. All human beings share these tendencies; they are a species-level problem. People in different cultures may react in different ways, but in the end, the tendencies are our common inheritance.

When I introduce the temptations in my workshops, people automatically feel uncomfortable, as if I have exposed something dark inside them that they normally do not like to admit about themselves. That instant shame response is another species-level obstacle that we have to overcome. Realizing that the temptations are something we all share by virtue of being human can take the embarrassment out of acknowledging and accepting them as part of who we are. Managing the fear of embarrassment is necessary in dealing with the temptations and in maintaining dignity when our instincts think they know better. Once we set embarrassment aside, we can learn how to control our hardwired instincts before they control us. Now let's find out what the temptations look like, how we can recognize the tug of our evolutionary legacy, and what we can do to resist the lure to violate our own and others' dignity.

# 11

# Taking the Bait

Don't take the bait. Don't let the bad behavior of others
determine your own. Restraint is the better part of dignity.
Don't justify getting even. Do *not* do unto others
as they do unto you if it will cause harm.

The subway terminal was packed. When the train pulled in, out
came another crowd of shoving, pushing people. One young man,
iPod plugged in, bumped into another man on the platform, nearly
knocking him over, but his victim was saved from falling by a woman
behind him. The young man with the iPod immediately said that
he was sorry, that he hadn't been paying attention, but the man he
had knocked into was so angry that he wanted to fight. He yelled
obscenities, and a couple of people with him held him back from
taking a swing.

I could see the fellow with the iPod struggling not to respond. He
took several deep breaths and stood motionless with his eyes fixed on
the would-be pugilist. I thought he was going to pounce. Instead, he
put the earbuds back in his ears, readjusted his backpack, and walked
away. The other man continued to scream at him as the train doors
behind him closed, the train pulled away, and the crowd dispersed.
I watched the victim's friends still attempting to calm him down

while the young man with the iPod walked up the exit ramp until he was out of sight.

What this story illustrates is the powerful temptation to lash back at others who have treated us badly. The reaction is a natural one: we are trying to protect ourselves under threatening circumstances.

The temptation to respond to a threat with a threat, to take the bait, as it were, may be an illustration of the downside of mirror neurons. Earlier in the book I described the discovery of these special cells in our brain that enable us to feel what another person is feeling. These neurons help us read the emotional experience of others. When someone else is feeling sad, these neurons automatically stimulate the same neurons in us, making us feel sad, too. It is a wonderful gift when these neurons enable us to feel compassion, to connect us with others in primal empathy. But the neurons also have the power to incite in ourselves the anger, hatred, and negativity that someone else is feeling.

What is immediately obvious in the story is that the threat to the man bumped into by the person with the iPod was relatively minor. His violent reaction could have spurred the iPod wearer to lash back, putting both of them in peril. But the iPod wearer did not take the bait. He fought his powerful instinct to fight back and won. And because he won, he could make a choice about how he wanted to behave. His I took charge, and he ended up with his dignity intact.

The response of the man who was bumped illustrates how our knee-jerk reactions are overkill today. When we let the threatening behavior of others get the best of us, our responses are often as bad as the initial provocations, but we don't see it that way because we feel justified in responding. The primal desire to get even blinds us to our own undignified behavior. At the time, and perhaps even later, because of lack of self-knowledge, we think it is okay to inflict harm on those who harm us. Indeed, we seek revenge and retaliation for perceived offenses in many subtle and often unconscious ways.

I accompanied a friend to the pharmacy to pick up a prescription. The pharmacist snapped at my friend when he asked her an innocent question about getting more than one refill at a time. "That's impossible. We're allowed to fill it only once. You'll have to come back."

It was her dismissive and negative tone, more than her words, that upset my friend. I could see his instant reaction. His face reddened and his voice tightened. He got angry and demanded to see her supervisor. The pharmacist said her supervisor wasn't there. I was beginning to get upset myself because my friend wouldn't let go of his anger. I tried to calm him down, but he acted as though I wasn't even there. When we finally left the store, we discussed what had happened. He admitted that he shouldn't have lashed back the way he did. I thought that was the end of the story.

I later found out that my friend had called the pharmacy the next day and had spoken to the supervisor, demanding that the woman be fired. Unfortunately, my friend was successful. The woman was asked to leave. It wasn't until many weeks later that I was able to hold a mirror up to my friend and get him to look into it. I pointed out that his desire for retaliation had cost this poor woman her job. I acknowledged that she had answered his question snippily, but I was able to get him to see that his behavior was equally undignified. He had taken the bait. Self-righteousness has the power to take over our best selves, compromising our ability to see how we justify harming others.

These examples show why it is so important to understand our powerful human instincts. Unless we have an awareness of our evolutionary legacy—the instinct that wants us to eliminate the source of a threat—we become a slave to it. But we have the capacity to catch ourselves when instincts take over our better judgment; we can halt and change our response. All of us have suffered myriad indignities in our lives; the pathway to the instinctive response is well traveled. Until we come to terms with our human vulnerabilities and know that we have the capacity, in the service of self-preservation, to do great harm to one another, we will be stuck in the never-ending cycle of indignity.

The better part of dignity is restraint. We are all able to hold ourselves back, to not take the bait. We just need to be aware that we have a choice. The other part of who we are—the I, as opposed to the Me—recognizes that we do harm to ourselves when we do harm to others. The power of self-knowledge is that we are able to use it when our instincts lead us into danger.

During times of conflict it is especially difficult to restrain our self-protective instincts. I see it in international conflicts. One group sends a suicide bomber into a busy civilian street, and the other side retaliates, killing scores of innocent people with bombs of its own. Our instincts want nothing more than to annihilate those perceived as the enemy.

According to a distinguished Pakistani diplomat, the anti-American sentiment that is still pervasive in the Middle East and the Muslim world generally can be attributed to the psychological indignities that the United States perpetrated in the aftermath of 9/11. In the attempt to restore damaged American dignity, the United States inflicted massive indignities on much of the Muslim world through its foreign policy. What is worse, our government felt justified in pursuing its policy.

Our Me, the home of our self-preservation instincts, knows nothing about solving problems, taking the perspective of the other, or feeling empathy. Unfortunately, only our I can see that our desire for retaliation, carried out full measure, would not only eliminate the other person but set up an endless cycle of violence. But by invoking the language of self-knowledge—of the I and the Me—we can identify who is running the show when we react to the bad behavior of others. Do we want our I or our Me to be making our decisions?

We do not need to let the bad behavior of others unconsciously determine how we behave toward them. This is where maintaining our dignity means using restraint when our self-protective instincts drive us toward rage and revenge. Violence and harm only provoke more violence and harm. It took my friend a while to see that he had taken the bait with the pharmacist. Yes, she had behaved badly by snapping at him. If he had recognized what was happening—that his instinct was to lash back when the pharmacist violated his dignity—he might have been able to catch himself. Instead, he entered a downward spiral of indignity, not only returning the harm but vastly increasing it.

After I shared with my friend some of the insights from the dignity model—how he could have restrained himself and not compromised his own dignity—he admitted that he had mixed feelings. Although his Me rejoiced, his I felt ashamed: a perfect example of what it looks like to accept the difficult truth of our complicated and often-embattled inner selves.

# 12

## Saving Face

*Don't succumb to the temptation to save face. Don't lie, cover up,*
*or deceive yourself. Tell the truth about what you have done.*

The temptation to save face is as powerful as our fight-or-flight instinct, but we may not be as aware of how automatic it is and how deep the impulse is to want to look good in the eyes of others. When we are confronted with a situation in which our hurtful words or actions are exposed, or in danger of being exposed, and we are not ready to admit to them, this instinct tells us to lie, to obscure the truth, to do whatever it takes to protect ourselves. The dread of having our inadequacy, incompetence, or lack of moral integrity made known is enough to turn us into liars, and we thereby violate our own dignity. Paradoxically, the lies are meant to protect our dignity, but the result is that we compound the self-inflicted violation.

How many times have we seen prominent and respected people lie about a wrongdoing when they are confronted with it? Former presidential candidate John Edwards went to extraordinary lengths to cover up the truth about having fathered a child with his mistress. He told a big lie, as if the bigger the lie, the better the chance that people would believe him. When the evidence of his indiscretion became incontrovertible, he publicly begged for forgiveness. He had been left with little choice but to be completely vulnerable. He ended up living out his worst fear.

Governor Mark Sanford of South Carolina committed the same error when it was revealed that he had traveled to Argentina to see his mistress. His first impulse was to cover up his extramarital affair: he declared that he had been hiking on the Appalachian Trail. When his lie was exposed by mounting evidence, he confessed on television that he had committed adultery. After such cover-ups, the public is often left with a feeling of disgust. If Sanford had been honest from the start and had admitted to the wrongdoing, the public might have offered compassion. Our mirror neurons can detect real feelings like genuine remorse and the desire to come clean.

We don't have to be a politician to fall prey to the temptation to save face. The desire to save face represents a direct link to our over-reactive self-preservation instincts; we are all vulnerable to it, since it is part of our shared humanity. People in some cultures are even more vulnerable than others—a Japanese participant in one of my workshops told the group that it is not uncommon for a Japanese man to commit suicide rather than face the shame of exposure.

When we make ourselves vulnerable by admitting the truth, the effect of the admission is often opposite from the one that our self-protective Me anticipated. Being vulnerable takes strength, and others are touched when they see our openness. Each of us knows the effort it takes to overcome the impulse to save face. If we are honest, we can see ourselves in those who are caught in the trap of deception, and we can hope that we would have the moral fortitude to do the right thing.

The struggle for dignity brings us face-to-face with our own internal battle between our I and our Me—the I wants to fight the fear of exposing the self and admit to our wrongdoing, and the Me wants to hide and fib. The struggle leaves us in a state of ambivalence until we acknowledge both personal truths, which frees us to make the right choice for the right reason—to preserve our dignity.

The human fear of humiliation is powerful, and what an obstacle it can be, not only to facing up to the truth about what we have done but to continuing our growth and development. Fears of looking stupid, incompetent, dishonest, or less-than in any way can keep us from moving beyond our current level of self-awareness; they can keep us enslaved by a variety of dysfunctions. Our relationships with others, as well as our dignity, are at stake. Tiger Woods's inability to deal with his dysfunctional behaviors with women damaged his marriage and family, not to mention his reputation. It took months before he owned up to his infidelity at a press conference.

The impulse to save face is as true for institutions as it is for individuals. Consider how long it took the Catholic church to admit to the cover-up of the rampant pedophilia among its priests, or the U.S. military to address the prisoner abuse at Abu Ghraib prison in Iraq. Neither churches nor governments want to look less than noble.

In evolutionary terms, exposing our misdeeds or admitting our mistakes leaves us vulnerable to being cast out of the group. When activated, the forces of self-defense instruct us to use any means necessary to save face, including lying, self-deception, and other forms of cover-up.[1] These powerful evolutionary forces create real obstacles to telling and facing the truth, making it difficult to resolve conflicts. When people feel that they have been wronged, they want acknowledgment of the wrongdoing; but those who do wrong are held back by these forces from telling the truth. This dynamic plays itself out at the political level as well as in our interpersonal relationships because it is a human, species-level phenomenon. It affects all conflicts in which human beings are involved.

Like all instincts to risk our own dignity in the service of self-preservation, the instinct to save face might protect us in the short run from the dreaded feeling of humiliation, but it doesn't serve our long-term interests. Instinctive responses do not factor in consequences or strategy. In *Stumbling on Happiness,* Daniel Gilbert makes the point that we consistently fail to accurately forecast how we are going to feel when we imagine the future.[2] That failure is a function of our brains. Our self-protective instincts are designed to react to an immediate perception of danger. When we sense that our dignity is on the line, when we get caught up in the emotional pull to save face, our long-term processing skills retreat, and dread avoidance takes the vanguard position. Once we understand the battles we face, we can develop the fortitude to fight them.

The best weapon for this fight is self-knowledge: knowing what we are up against and knowing that we are capable of so much more than responding by instinct. We have the power to overcome the preprogrammed responses. Doing so is difficult but not impossible. We have to work at it. We have to choose to learn how to manage our reactions.

But choosing is not like trying to decide between chocolate or vanilla ice cream. Rather, we come face-to-face with the reality of our humanness. We must realize that these instinctive responses can put our dignity in jeopardy, and we must fight the temptation to protect ourselves. In the end, the choice is not to save face but to save our dignity. To me, that we are left to make that decision for ourselves, that the dignified choice doesn't come naturally and requires an internal struggle on our part, is at the heart of the challenge of our continued development.

# 13

## Shirking Responsibility

Don't shirk responsibility when you have violated the dignity of others. Admit it when you make a mistake, and apologize if you hurt someone.

The temptation for us to shirk our responsibility when we hurt others is similar to the instinctive reaction to save face. We prefer to minimize a painful incident and hope that the memory of it will go away. Sweeping the issue under the carpet is the choice of our self-protective Me. That part of us would rather deny the wrongdoing than have us be held responsible and have to change our behavior.

How do we overcome this powerful impulse? How do we shift this internal default setting from defensiveness to accountability when the situation clearly calls for it? How do we rise above our inclination for self-preservation when the self that we are preserving is in desperate need of change?

During the first hour of a dialogue workshop that my colleagues and I had organized for students from the Middle East living in the Boston area, a young Palestinian student (I will call her Rahima) hardly spoke. While her Palestinian friend and an Israeli student whom I will call Don argued about the way Palestinians were treated when they attempted to make the border crossing between the West Bank and Israel, Rahima sat nearly motionless as she listened; only her eyes shifted back and forth between the interlocutors. Her Palestinian friend was getting frustrated. He said to the Israeli, "If you think we're not treated like dirt at the checkpoints, dress up as an Arab and try making the crossing yourself."

The argument continued, getting nowhere. My colleague, who was chairing the session, was just about to intervene when Rahima interrupted the men and asked to speak. The room went silent. She looked at Don, the Israeli, and without an ounce of judgment in her voice said to him, "I can see that you are having trouble believing my colleague. Let me tell you a story that might help you understand what he is trying to convey.

"When I was about six years old, my grandfather told me that we were going into the Old City of Jerusalem to visit a friend whom he hadn't seen in many years. I remember thinking at the time that my grandfather was old, and I wondered if he wanted to go to see his friend in order to say good-bye. This thought made me sad, because I loved my grandfather, yet I was excited to accompany him on the journey.

"He was a prominent member of our community in Ramallah; everyone respected him. People young and old would come to him for advice—he was the unofficial mediator. I was proud to be his granddaughter.

"When we approached the border crossing, a young Israeli soldier asked my grandfather to get out of the car. I was terrified. The soldier was carrying a big gun, and I didn't know what was happening. At one point, I saw my grandfather trying to explain something to the soldier, but the soldier accused him of lying and started yelling at him. I couldn't believe it. I jumped out of the car, went up to the soldier and said to him, 'What are you doing? Don't you know who he is? He's my grandfather! You can't talk to him that way.'"

The room went silent. Rahima put her face in her hands and sobbed. We waited.

Don was the first to speak. He turned to Rahima and told her how sorry he was that she and her grandfather had had to endure that humiliation. His voice was trembling as he explained to her how difficult it was for him to take in her story. "As an Israeli, I believe in my heart that we are good people, fighting a painful war that we have to fight in order to maintain our Jewish identity and a future for the Jewish people. I feel the righteousness of our cause. If I accept what you say, that we are going about this fight in a way that is not righteous, in a way that is profoundly harming you and your people, then it forces me to look at my own identity and say, 'Who am I? What am I doing?' I cannot take in your experience and keep my sense of who I am intact at the same time. Because I now have no doubt that what you say is true, I have to swallow a bitter truth: that the way I have constructed my identity up to this point is causing great suffering for the Palestinian people."

In all the dialogues I had facilitated before and in all I have facilitated since, I have never witnessed such bravery. Don made himself vulnerable in front of his fellow Israelis, in front of the Palestinians at the table, and in front of the rest of the participants. We were all speechless. There was nothing left to say. Words do not belong in such moments of reverence. We closed the session, and the Palestinian who had been arguing with Don earlier walked up to him and shook his hand. They gathered their belongings and quietly left the room together.

Nearly twenty years have passed since that session. Fortunately, it happened early in my career, and I have called upon the memory many times to refresh my belief in what is humanly possible.

Some people in my business argue that identity is sacrosanct, so asking people to change aspects of who they think they are is un-warranted—and in any case, they say, it is impossible. I have never believed that, although our human nature does make it very difficult for us to change. Parts of our evolutionary legacy are geared to preserve the self even when the self is behaving in ways that are self-detrimental.

Taking responsibility for the harm done to others and oneself is not part of the evolutionary agenda. Survival is all that matters. Consideration of the right thing to do is left to the I, the part of us that makes us uniquely human. Steven Pinker reminds us that evolution, with its agenda to keep us alive and reproducing, does not have a conscience.[1] But we do. We have what it takes to become aware of the ways we are undermining not just others by being enslaved by our instincts but our own growth and development.

Under normal conditions our identities evolve and change just like any other aspect of human development.[2] By "normal conditions" I mean an environment and relationships that support and encourage growth. But if the environment is hostile, the mechanisms that allow the healthy evolution of one's sense of self shut down in the service of self-preservation.

One's identity requires a sense of safety and stability to evolve. Lacking safety and stability, our sense of who we are becomes frozen in time.[3] It is too threatening to open oneself up to the possibility of change and would be far too psychologically disruptive. We need a stable inner base from which to ward off threats. There is no time to consider the effects of our actions on others when we are defending ourselves.

When threats to our well-being are the norm, is it any wonder that we find it difficult to examine ourselves with improvement in mind and to open ourselves to new learning? The way we make sense of the world gets stuck in survival mode. Since looking at the contribution we might be making to a failed relationship would be destabilizing, deflecting responsibility appears to be a default reaction. Deflection gives rise to a number of cognitive distortions about the other, further limiting our capacity to self-reflect.[4] Deflecting blame and responsibility becomes the operative response, putting the normal development of our identity on hold indefinitely.

What I found so remarkable about Don's intervention during that Harvard workshop was that he overcame the self-protective impulse to guard his frozen identity. In front of us all, he opened himself up to taking in painful information that contradicted who he thought he was. The emotional force of Rahima's story was more powerful than his defenses. It is worth noting that the whole time she told her story, she was looking at Don. I didn't know it then, but now that I have a deeper understanding of how humans are hardwired to connect, I realize that what happened between them was a restoration of their capacity to empathize with one another.[5] Their mirror neurons were reactivated and in synch. The way Rahima recounted the story—as if she were right back there in the moment—helped Don feel its reality.

The Palestinians in the room also opened to Don. The frozen aspects of their identity—the parts that wanted to see Israelis as all bad and all to blame—also melted a bit. The melting on both sides restored much more than the potential to restart their individual processes of identity transformation—it restored the humanity to the Israeli-Palestinian relationship, at least in that room.

When Rahima told her story and Don revealed his feelings, the destructive dynamics in the room shifted quickly, in a matter of seconds. But the overwhelming feeling I was left with was how fragile our identities are and, at the same time, how rigid they can become in the face of threats. "Frozen" is an apt metaphor. Without the capacity to feel empathy with another person, the warmth that normally holds us together dissipates, and we freeze into our isolation.

Once the channels of empathy opened up between Don and Rahima, once he experienced what she had felt as a little girl watching her beloved grandfather being treated badly, he began to care about her. After she told her story in such a moving way, he could no longer deny her suffering or the role his community had played in it. He was emotionally forced to come to terms with an aspect of himself that he had not looked at. But because he cared about her in that moment, he had to open his eyes.

The connection formed at that workshop reminded me of a talk by Kathy Roth-Douquet, which I heard at a conference at Harvard Divinity School. She made the simple but profound point that we need to build strong relationships in which people care about one another because "when we care, we learn."[6] In this case, probably in all cases, what we need above all when we are embattled is to create the conditions for caring to emerge, and we can't do that just by cognitively understanding the perspective of the others. Caring is an emotional event that allows us to resume the developmental process of taking in important information as we interact with the world and expanding our sense of who we are. When hostility is replaced by genuine caring, we can better tolerate information from others about how we are causing them harm, enabling us to look at ourselves with new eyes.

*When we care, we learn*—this insight can perhaps be applied to the story of Don and Rahima. The power of genuine human connection can overcome our self-protective instincts even in the most damaged relationships. Learning how we have harmed others is the prelude to the next compelling insight: *Once we are aware, we are responsible.*

If we inflict dignity violations on others unknowingly, it is easy to deflect responsibility for what we have done. But once we learn what it means to have violated others, we can no longer use ignorance as an excuse. Being responsible means that we hold ourselves accountable for our actions, especially those that are hurtful to others.

# 14

## Seeking False Dignity

*Beware of the desire for external recognition in the form of approval and praise. If we depend on others alone for validation of our worth, we are seeking false dignity. Authentic dignity resides within us. Don't be lured by false dignity.*

*Scenario One.* I have a friend (I will call her Maura) who has frequent bouts of depression. In a vulnerable moment, she once told me that she never feels as though she measures up; she has a hard time seeing anything but her imperfections, and her internal dialogue is fraught with self-doubt. Looking at her from the outside, I find it hard to imagine why. She is smart and attractive, has a high-status job, and makes a six-figure salary.

True, her personal life was a mess. She attributes her difficulty in maintaining a relationship to her demanding job. When I asked her why she worked such long hours, she confessed that the office was the only place where she felt good about herself. Much of her sense of well-being derives from her status as a businesswoman. She was terrified at the thought of losing her job—the primary source of her sense of worth and well-being.

*Scenario Two.* Many of the successful men I know have a common refrain: "I've made it. I have all the money I need, I can go where I want, I do what I want, and I don't have to answer to anyone." One friend, whom I will call Jerry, confessed to me that despite rating high on all the external measures of success, he still doesn't feel at peace with himself. He often feels inferior to others and is afraid that people will eventually see him for who he really is: weak, vulnerable, and fearful. He said that he often feels uncomfortable in social situations, even though others find him funny, engaging, and delightful to be with. He has to work very hard to relax with people, especially those he looks up to.

*Scenario Three.* Mae and Lavinia (not their real names) met during their first year at an all-girl boarding school in Massachusetts. They were both fourteen years old, international students from Hong Kong, and in the same grade. Lavinia comes from a family of high-powered architects, lawyers, and Ivy League graduates—an elite family, especially in Hong Kong, where class divisions are clearly understood. Mae, on the other hand, comes from a family of blue-collar workers and farmers; she was the first in her family to go to college.

Many class-related dignity violations arose in the decade of interaction between Mae and Lavinia. According to Mae, Lavinia's high status in Hong Kong society meant that she was raised to believe that she was better than any ordinary worker. Her attitude of superiority dominated many of their interactions, often leaving Mae feeling humiliated in front of others.

One of Lavinia's favorite ways of putting Mae down was to make fun of her Chinese accent when they were together with American friends. Lavinia, who had been raised by a British nanny, spoke excellent English. She joked about Mae's farmer family, pointing out that her own father wouldn't think of getting his hands dirty; he left that kind of work to people who were ignorant and uneducated. Even though Mae did very well academically, her relationship with Lavinia plagued her. Was she doomed to feel inferior all her life because of her class background?

❀

These three scenarios show how we can be tempted by false dignity to violate our true dignity. False dignity is the belief that our worthiness comes from external sources. It takes many forms. If we need praise and approval from others in order to feel good, if we seek high-status positions to prove that we are successful, if we believe that some people are superior simply because they were born into a particular class, race, or ethnic group, we hold a false sense of dignity. When we think that our self-worth comes from outside ourselves, we are in danger of seeking false dignity. When we lose sight of the fact that we are inherently valuable, that our worth is not dependent on external validation, and that we matter as human beings, we let our true dignity slip out of our hands. When we let our Me dominate our consciousness and forsake the stability and grounding that only our I can give us, our dignity is in jeopardy. False dignity comes and goes if we are dependent on approval, praise, and recognition from others to feel worthy.

True dignity comes with our full acceptance of the miracle of our existence. The real question is, Why do we lose sight of this important truth about who we are—that every one of us is worthy of being treated well? Why are so many of us vulnerable to seeking false dignity?

I have come to the conclusion that at this moment in human history, most of us have yet to develop an awareness and acceptance of our emotional strengths and vulnerabilities, and most of us have yet to realize the extent to which those strengths and vulnerabilities rule our inner worlds. We have not taken up the call to "know thyself"—the admonition inscribed on the wall of the Temple of Apollo, the Greek god of wisdom, in Delphi, where the oracle made her prophecies.

Because of our usually stunted emotional development, most of us have self-preservation instincts that are unchecked and that have therefore caused many dignity violations; because we don't understand our reactions, we have little option but to retaliate or retreat when someone harms us. Our ignorance has caused a lot of suffering. And since we do not know how to nurture and honor our own inherent value, we live in a state of inner turmoil.

Many of us lost touch with our inherent worth because we were not treated with dignity as children. Our caretakers did not know the importance of honoring dignity or the lasting effects of not honoring it.

Specialists in child development now know that for children to gain a sense of inherent worth, they need to have it mirrored back to them early on—in fact, as soon as they leave the womb.[1] They need to see the joy and love in their caretakers' eyes when their caretakers look at them. Their felt experience of worth is built up every time a caretaker pays attention to them, smiles and coos at them, responds when they cry. They need loving attention as much as they need attention to their physical well-being. This is a constant requirement throughout a child's life. It takes on different forms as the child progresses developmentally, but as a general rule, love is attention.[2] What love looks like, in my view, is like treating others with dignity; it is far more than a feeling.

If we have "good enough" mirroring as children, the connection to our I develops.[3] Children who experience good-enough mirroring don't grow up with crippling self-doubt.

Although I know people who have a strong I and are aware of their inherent worth, they are the exception. As a species, we have little awareness of the importance of honoring dignity; some of us are still looking for the love and approval in others' eyes that we needed early on. We have developed a Me-dominant consciousness; we believe that there is a deed to accomplish or a quality to acquire before we can feel good about ourselves. This lack of a strong I in adulthood because it lacked childhood nurturing is a paradox: our I is part of our birthright, but it needs to be strengthened early on to become the stabilizing source of our dignity.

Maura was lured by false dignity—the belief that her job performance was the only source of good feelings about herself. After getting help from a good therapist, she realized that what she really wanted was acknowledgment and recognition. Her childlike thinking about how to get acknowledgment stemmed from her futile efforts to elicit the attention of her workaholic father, who spent little time with her when she was growing up. What was especially confusing for her was that he always told her he loved her. But remember, love is attention. Neglect, for whatever reason, is not love. What she needed was for him to spend time with her. Because he was never around, her childlike way of making sense of the world made her think that she had to win his attention through extraordinary achievements. The belief that she had to excel in order to feel good arose long before she knew what she was doing and why.

My friend Jerry was also caught up in false dignity. All his life, his Me drove his quest for dignity through success, only for him to discover that he was looking in the wrong place. He didn't find dignity in his multimillion-dollar bank account. It shouldn't be surprising to learn that he, too, had a childhood fraught with dignity violations.

Jerry told me one of the expressions that he heard a great deal during his childhood was "Children should be seen and not heard."

So many of the elements of dignity are violated by that saying. Children (like all human beings) need to be listened to, acknowledged, understood, and responded to. If they are silenced, they are not able to communicate what they are experiencing. They learn that their feelings and experiences don't matter. For them, that translates into "I don't matter." And when the I doesn't matter, nothing is left but the Me. Jerry's Me is still scrambling for dignity. Again, love is attention.

This is not to say that children should be allowed to do whatever they want. In my view, permitting children free rein, allowing them to dominate the lives of their parents, is just as dangerous as neglect. Discipline plays as big a role in educating about dignity as being seen, responded to, listened to, and understood. Children also need to learn restraint. The temptations they experience are raw and primal. We do them no favors by giving them whatever they want. They run the risk of becoming narcissistic adults who are incapable of self-criticism, much less self-restraint.

In the third scenario, both Lavinia and Mae located their sense of worth outside themselves. Lavinia was using her class status to feel superior to Mae. Mae, on the other hand, accepted the false notion that because of the circumstances of her birth—her lower social class—she was inferior. Both succumbed to the temptation to believe that their dignity, or lack of it, could be explained by their social status.

Besides the temptation of false dignity, the three scenarios have one other thing in common. None of the people is happy in relationships. Maura and Jerry are each longing for a life partner, but they can't seem to find one, and when they do meet someone interesting, they always end the relationship. Mae, in contrast, stays in relationships that are not healthy.

Our evolutionary legacy is partly responsible. As I have said, we are hardwired to respond to a threat to our well-being—we either fight or flee. These reactions have helped to ensure our individual survival and the survival of our genetic material into the next generation. But we are also hardwired to be in connection with others, because being in relationships also helped our early ancestors survive. People were better at warding off threats in a group than alone.

❀

Our evolutionary legacy thus has the potential to wreak havoc on our relationships. What we need for survival (a connection to the group) can also feel like the greatest source of threat to our survival (when the connection becomes hurtful). These two realities can live comfortably side by side when we are getting along with one another. The problem arises when we threaten each other's dignity—when a relationship becomes the source of pain instead of pleasure.

This dramatic tension between the primal desire to be connected to others and the hardwired instinct to protect ourselves from others' possibly hurtful attacks explains why relationships are so hard. In Elizabeth Gilbert's book *Committed,* she describes her relationship with her new husband this way: "Filipe and I had each learned firsthand this distressing truth: that every intimacy carries, secreted somewhere below its initial lovely surfaces, the ever-coiled makings of complete catastrophe."[4]

Our sensitivity and vulnerability to being injured by others and our ability to injure them—both part of our evolutionary legacy—set us up for struggles in our relationships. Relationships break down when the need for individual self-protection overrides the need for connection. We walk away from a relationship—political or personal—if it begins to feel threatening or does not serve our individual desire to be treated well. Our self-preservation instincts predispose us to favor disconnection over connection: "better safe than sorry." Is it any wonder that intimacy feels threatening? Being vulnerable sets us up for being injured or re-injured.

The dignity of individuals also is compromised when their need for connection overrides their need for self-protection. How many times have we compromised our own dignity—by, say, not reacting to being violated—for the sake of maintaining a relationship?

For both Maura and Jerry, their need for self-protection was more powerful than their need for connection. They both had similar early experiences of not being seen, acknowledged, recognized, or heard. They never learned that being in a relationship was a source of pleasure and mutual understanding. They were preoccupied with protecting their early wounds and worried about being injured again.

Mae's case is different. Her early experiences positioned her to stay in the relationship with Lavinia, even though it was painful. Her internalized belief in her inferiority convinced her that she could gain dignity by befriending someone from the "superior" class. She stayed in the relationship rather than risk the loss of (false) dignity that she derived from her friendship with Lavinia.

The lure of false dignity is powerful. The short-term gain—the wonderful feeling we get from being praised or acknowledged—is deeply gratifying, if not addictive. We expend precious energy satisfying our Me, where that insatiable desire for recognition resides, or worrying about not satisfying it. Without the balance of our stabilizing, unconditionally worthy I, we cannot be in harmony ourselves or in harmony with the rest of the world.

# Seeking False Security

Don't let your need for connection compromise your dignity.
If we remain in a relationship in which our dignity is routinely violated,
our desire for connection has outweighed our need to maintain
our own dignity. Resist the temptation to settle for false security.

Maria (not her real name), a participant in a dignity workshop, had been trapped in a relationship with her former husband, Bill (not his real name), for a number of years before she realized that her fear of not being able to manage on her own was not only keeping her in a dysfunctional relationship but also holding her back from fulfilling some of her own aspirations and goals for her life. Bill had been having an affair with another woman for years, but that was only part of the problem in their marriage. Once Maria learned the essential elements of dignity, she was able to articulate the many ways that Bill was violating her dignity, one of which was especially painful. In the heat of their arguments, he often mocked her identity (she was a Mexican American), claiming that she would never be a "real" American. This was especially hurtful because she often felt like an outsider. His comments exacerbated her feeling that she didn't belong. She admitted that she could dish back the cruelty, however. She often violated his dignity as well—she took the bait.

Why didn't she divorce Bill? Because she didn't feel safe enough to make it on her own. Ironically, what Maria feared the most was the sense of insecurity that she imagined she would feel without Bill, but the constant violations of her security in the marriage were not enough to motivate her to leave him. Not until her mother came for a visit did Maria finally wake up to how she was violating her own dignity by remaining in the relationship. The sense of security that she was clinging to was false security. In fact, she was compromising her own psychological safety by not taking action to protect herself from further emotional abuse.

A central theme in Daniel Goleman's book *Social Intelligence: The New Science of Human Relationships* is that human beings are wired to connect. He explains that our brains are equipped with neural mechanisms (mirror neurons) that enable us to be in synch with one another, to feel each other's emotions in a way that affects us both positively and negatively. When we are with someone whom we care deeply about, the feeling is usually pleasant. But the neural mechanisms are equally responsible for making us feel bad in the presence of others, inciting a cascade of negative emotions. Either way, whether we feel positively or negatively toward someone, we are connected. Our brains are wired in myriad ways to attune ourselves to one another, affecting not only how we feel in any given moment but our health in general: "Our brain-to-brain link allows our strongest relationships to shape us on matters as benign as whether we laugh at the same jokes or as profound as which genes are (or are not) activated in T-cells, the immune system's foot soldiers in the constant battle against invading bacteria and virus[es]."[1]

Goleman explains how this interpersonal connection can both hurt and help. If a relationship is healthy and nourishing, the partners will contribute to each other's health and well-being. But if the relationship is not healthy, one partner can introduce toxicity into the other's system. Goleman feels that the most important insight to come out of the new field of social neuroscience is that "the social brain represents the only biological system in our bodies that continually attunes us to, and in turn becomes influenced by, the internal state of people we're with."[2]

Because of our innate capacity for connection, the effect we have on one another's internal states cannot be overstated. We can add to this new understanding the discovery that our brains have tremendous "neuroplasticity." In other words, if we are exposed to a negative relationship for long periods of time, the exposure can actually reshape our brains. As Goleman says, "How we connect with others has unimagined significance."[3]

These findings shed light on why it feels so difficult, as it was for Maria, to leave a relationship in which dignity is repeatedly violated. Maria's brain had become inured to the poisonous effects that her husband's behavior had on her; she failed to recognize that by staying in the relationship, she was inflicting damage on herself by allowing the abusive behavior.

With the help of a therapist and the support of friends, Maria was able to extricate herself from her marriage. She reclaimed her dignity as well as her understanding of what real security meant for her. In her words, "Any relationship in which your dignity is being routinely violated is not a safe relationship, no matter how much you deceive yourself into thinking otherwise."

An even bigger point, for the purpose of this book, can be made about the biological reality of the effect we have on one another. When we become aware of the power we have to influence others in a positive way, we can make a significant difference not only in our lives but in the lives of those we come in contact with. The research tells us, beyond the shadow of a doubt, that the way we treat one another matters. The question is, What does it look like to treat others in a way that contributes to their health and well-being? The answer: *It looks like honoring their dignity.* The ten essential elements of dignity provide ten specific ways to honor dignity. They also provide an understanding of the opposite—ways to have a negative impact on others.

Once we understand our biology and the profound effect we have on one another, I don't believe we can approach encounters with others without thinking about it. This knowledge has the potential to change how we choose to live our lives and what responsibility we have toward others. Let me quote Goleman again: "The biological influence passing from person to person suggests a new dimension of a life well lived: conducting ourselves in ways that are beneficial even at a subtle level for those with whom we connect. Relationships themselves take on new meaning, and so we need to think about them in a radically different way. The implications are of more than passing theoretical interest: they compel us to reevaluate how we live our lives."[4]

Although we might all agree with Kant's moral imperative to treat people as ends in themselves and never as mere means, neuroscience has taken the issue far beyond morality, turning the matter of how we treat one another into a biological necessity. Honoring dignity is not just about being nice. It even transcends the issue of survival. Instead, it is about living our lives in a way that promotes each other's physical, psychological, and spiritual well-being and expands our humanness.

## Avoiding Conflict

Stand up for yourself. Don't avoid confrontation when your dignity
is violated. Take action. A violation is a signal that something
in a relationship needs to change.

To speak up when someone has violated our dignity is hard to do. How often do we let everyday violations slide by and say nothing about them? Let's say a colleague dismisses a concern you raised about whether to move ahead on a project. A majority of coworkers would like to call a halt. But without seeking to understand your perspective, the colleague continues to talk about moving the project forward. Or, you are in a staff meeting and your boss announces that he is planning to override a decision you made about retaining an employee who had repeatedly harassed a number of people. After weeks of careful investigation, you had come to the conclusion that the employee should be asked to leave. You are incredulous that your boss didn't even talk to you about your decision before overturning it. Or, your spouse screams at you in front of your children—which you have asked your spouse a million times not to do.

When I am faced with these kinds of situations, I often make excuses for keeping quiet: "It's not worth discussing—he'll never change," or "It's not a big deal," or "How much does the relationship really matter to me anyway? Do I really want to have a possibly unpleasant confrontation?"

We often resist bringing up offenses that are likely to cause a conflict with others, and no wonder. We resist taking feedback from others, and we fear that if our words might be taken as criticism, we will make ourselves vulnerable to the vengeful attacks that might follow. This is another example of how some of our outdated survival instincts have set us up for problems in our relationships.

When we let the hurtful actions of others pass, we might be avoiding conflict, but we are not avoiding injuries. We are setting ourselves up to be repeatedly violated, but at what cost? What we overlook is that our dignity is at stake. By letting our desire to avoid confrontation take over, we are giving others permission to do us harm.

Confronting people who violate us is indeed difficult, especially when the consequences could be serious: you might lose your job, a friendship, a business partner, or a spouse. And there may be times when we choose—wisely—to let a violation go by without saying anything to the offender. If the offender (such as a boss) has power over us, our reluctance to confront him or her is due to more than our evolutionary inheritance; there are serious practical consequences. Do we have other options?

The all-or-nothing approach—say nothing and suffer silently, or speak up and risk a loss—shows how loss avoidance plays as big a role in this temptation as conflict avoidance. It illustrates the delicate balance between our desire for stability and our desire for change. What underlies both of these fears of loss is emotionally volatile, putting us potentially in harm's way, so our instincts tell us to protect ourselves, even if our dignity is on the line. This survival reasoning blinds us to alternatives. There are some.

Before going into the options, let's go back to a point made earlier.

When we interact with other people, our social brains are building a neural bridge between us. The neural circuits activate our emotions, and the more frequently we are exposed to others, the greater the emotional force that gets activated.[1]

The flood of emotional energy is good when what we feel with others is positive and life-enhancing. The hormones and other chemicals released when we are enjoying an interaction have been shown to have a beneficial effect on our health. In *Love and Survival,* a world-renowned physician, Dean Ornish, reports that even severe heart disease can be reversed with lifestyle changes.[2] The lifestyle changes that he suggests do include diet and exercise, but the more startling results point to the role that loving, positive relationships play in the reversal of disease. When we are treated well, we heal.

Problems occur when we have repeated encounters with people who have a negative effect on us. Daniel Goleman describes research showing that our immune system eventually becomes compromised with the "slow poison" that is released with long-term exposure to harmful people.[3]

Once we know the effects of relationships, we have to ask, "How much do I want to expose myself to people who repeatedly violate my dignity, putting my physical as well as my mental health at risk?" The reality of the biological effects also forces us to come to terms with our evolutionary legacy. Acknowledging and managing our impulse to back away from confronting people who harm us gives us room to consider other choices. A story about Laura, a participant in one of my workshops, provides an example of the alternatives that the dignity model offers (the names in the story are not the participants' real names).

❉

While having lunch in the office cafeteria, Laura told her friend and colleague Eric that she was seriously considering confronting their boss, Mark, about an incident that had taken place the day before.

"Laura, are you sure you want to do that?" Eric asked. "It's risky business. What if he fires you on the spot? Even if he doesn't do that, he can make life miserable for you here."

The previous afternoon Laura's assistant, Rachel, had been putting together the final pages of a report that Laura had written and that Mark needed by the end of the day. Rachel was sitting at her computer, with her back facing the hallway, when she realized that Mark was peering over her head, reading the page of the report that was on the monitor. Before she could speak, he said, "What's wrong with you? Don't you know how to format a report that goes to the dean's office? You've done it all wrong! Send the document to my assistant. She'll do it right. You know, Rachel, in this economy, we can get people with PhDs to do jobs like yours. Take this as a warning."

He didn't give her a chance to explain that she was working on a draft and that she had every intention of using the proper format in the final report. As hurtful as Mark's comments were, the violation was compounded by his biting tone. On top of that, two people standing in the hallway nearby couldn't avoid hearing Mark's punitive remarks. Rachel felt publicly humiliated.

Mark had been the topic of many lunchtime conversations among his employees, all of whom had suffered from his indignities and all of whom were afraid to confront him. Speaking up would have been career suicide, or so most of them believed. They needed their jobs and felt resigned to Mark's behavior: "It's just the way he is."

Laura had struggled with Mark's behavior ever since she took the job six months before. Mark had embarrassed her in staff meetings with inappropriate jokes and had often questioned her judgment in front of others. His caustic manner in meetings made the environment next to unbearable.

Laura felt sick just thinking about confronting Mark. She felt even worse knowing how bad Rachel felt; Rachel had broken into tears when recounting what had happened. As Rachel's boss, she believed that she had an obligation to protect Rachel. She was struggling with what felt to her like a moral dilemma.

Looking in from the outside, we can see that Laura had to say something to Mark. By not calling him on his undignified behavior, she, and all the rest of the staff, were exposing themselves to daily dignity violations, not to mention playing a big role in enabling him to continue to cause harm. Laura knew she was in a better position than anyone else to talk to Mark because he favored her, but she still dreaded the confrontation.

Fortunately, Laura realized that by not taking action on Rachel's behalf, she would be shirking her duty as her boss. By not defending her assistant's dignity, she would be compromising her own. What were Laura's options?

❋

*Option One: Speak Up.* Before confronting Mark, Laura must be clear in her intentions: "I do not want to inflict harm on him; rather, I want to give him feedback about the ways he is harming others. He may not be aware he is causing harm." Being sincere in this intention requires self-awareness. Laura needs to accept that a part of her (the Me) does want to return the harm. Remember, the Me wants to eliminate the source of the threat; the desire to fight is real and difficult to ignore. To overcome it, Laura must acknowledge that it exists and recognize that if she allows that part of her to direct the conversation, she is likely not only to return the dignity violation by inflicting harm on Mark but also to violate her own dignity in the process.

To give Mark feedback about his behavior that he might be able to take in, she needs to be situated in her I, the part of her that is capable of seeing beyond her instinctive and understandable desire to lash out. She needs to be aware of her own complex humanity (since we are so much more than our hardwired instincts) as well as his. She should talk to friends and practice what she is going to say. She needs to keep herself calm and let her I do the talking.

Besides being clear in her own intentions, Laura needs to grant Mark his dignity. She needs to stay focused on the fact that Mark is a human being with inherent worth, value, and vulnerability. Her task is to separate Mark as a human being, worthy of dignity, from the way he is behaving. She needs to give him the benefit of the doubt that he is capable of hearing what she has to say.

Once Laura has prepared for the confrontation, she needs to gauge Mark's capacity for empathy. Does she have evidence that Mark is capable of feeling for other people, that he cares about how his behavior affects others? If so, then the chances are good that talking to him about his behavior may lead him to examine it and make changes. Even if she is unsure, it is better to assume that he does care about his impact on others—it is better to give him the benefit of the doubt.

If, on the other hand, Laura has evidence that he is not concerned about the effect he has on others, that he might have narcissistic tendencies that make it unlikely that he can see others as separate from himself, then the situation is trickier. According to Alexander Lowen, the author of *Narcissism: Denial of the True Self*, narcissism is the result of a distortion of normal self-development brought on by the failure of parents to provide sufficient nurturing and emotional support.[4] The parents (or caretakers) failed to recognize and respect their child's individuality and maybe attempted to mold the child in a particular way. A real self did not emerge for the child; instead, there was an image of an ideal self—usually one of perfection—and a desire on the parents' part for others to see the child in the prescribed way. The investment in the image, as opposed to a true sense of self, characterizes a narcissist. Energy is directed toward the enhancement of the image at the expense of the narcissist's true feelings.

Without the experience of being seen, responded to, and valued as an individual of intrinsic value, the child does not develop an internalized sense of worthiness. A sense of dignity—false dignity—comes from the external image rather than from the self itself.

For narcissists, criticism is intolerable. Without an internalized sense of self-worth that they can anchor themselves to when others give them feedback about their behavior, even constructive comments are taken as negative—blows to the perfect image they depend on for their sense of worthiness.

Even when people have developed an internal sense of self and an awareness of their own inherent value, having blind spots pointed out causes discomfort. It is wrenching to hear that you are violating the dignity of others, even or especially if you are not aware of it. Because most of us do care about how we affect others, being told that we have (unknowingly) hurt them is hard to digest.

When we learn about our blind spots, we inevitably experience what I talked about earlier as a tolerable level of shame. Shame is impossible to avoid for those of us who are aware of and accept our dignity. In *Emotions and Violence,* Thomas Scheff and Suzanne Retzinger describe this functional aspect of shame as important in keeping relationships together. They also point out that there is pathological shame—excessive, intolerable shame—that can break up a relationship.[5] Unless we feel bad about what we have done, we may be disinclined to do the demanding work of changing our behavior to maintain a relationship; there might be, instead, an inclination to blame the other person involved. But a tolerable level of shame gives us the motivation to act on the new awareness and not repeat the behavior again in the future.

Because narcissists have an unusually difficult time taking feedback, it is difficult to give, no matter how it is delivered. Even separating the feedback from the person receiving it, reminding the narcissist that the deed and the doer are not the same, will fall on deaf ears. By holding up a mirror to a narcissist, you are playing out his or her worst fear: exposing the narcissist's inadequacy and worthlessness to the world. Without an I that is grounded in true dignity, the Me-dominated narcissist, whose dignity comes entirely from external sources, will be unable to bear looking bad in the eyes of others. The shame is intolerable.

If Laura has decided that Mark is capable of empathy and has a strong enough sense of his own dignity to be able to tolerate the inevitable shame associated with what she will say, then her chances are good that she will be successful.

In this scenario, Laura decided that Mark was capable of empathy and that he did care about how he affected people. That conclusion was problematic to reach for two reasons: her dread of his reaction and her belief that he was incapable of change. Taken together, these beliefs generally make it difficult for us to speak up. Because Laura knew the dignity model and knew how her evolutionary instinct to protect herself would tempt her from doing what she felt was right, she was able both to overcome her dread and to give Mark the benefit of the doubt. Hoping that he was capable of hearing what she had to say, she screwed up her courage and asked him if he had a minute to talk with her. He welcomed her and asked her to sit down. Here is how the conversation went.

Mark:  So, Laura, what can I do for you?

Laura:  Oh, I just wanted to talk to you about something before too much time passed.

Mark, I want you to know that what I'm about to say is difficult for me, but my relationship with you is important to me, and I feel that if I don't talk to you about what happened yesterday between you and Rachel, it might get in the way of our ability to work well together. First, I want you to know that I respect you as my manager, and I also know that you are trying to do the best job you can.

Mark (*shifting in his chair*):  Boy, this doesn't sound good.

Laura:  No, Mark. Just the opposite; it really is good. If I didn't care about you, I'd just let this pass. But because I really believe that you are not aware of how you come across sometimes with people, I thought that as a concerned friend and colleague, I would try to help you see something that I am convinced that you don't see.

Mark (*sounding a little agitated*):  Okay, so what is it?

Laura:  Yesterday when you spoke to Rachel about the report she was preparing, she said that you assumed that she was working on the final copy and thought she hadn't formatted the report in the right way.

Mark:  You're not going to make a big deal about that, are you?

Laura:  Well, it was a big deal to Rachel. She felt embarrassed by the way you talked down to her, assuming the worst. And then when you implied that she could lose her job, she was really upset. My guess is that you were not aware that other people were standing in the hall-way, and they overheard the whole thing.

Mark (*visibly upset*): This is ridiculous. My assistant would have had that report done a lot faster than Rachel. She really isn't all that good. Now I'm not sure I want to continue with this conversation.

Laura: Mark, I can completely understand your reaction. I'm sure I would get upset, too, if my subordinate came into my office and said what I said. What I'd like to make sure of before I leave is that you heard what I said about my reason for coming in here: that I care about you and I don't want anything to interfere with our good working relationship. And maybe—even more important—that I am sure that you are not aware of how you come across at times. My husband points out my blind spots all the time. The fact is, we're all human and we all have things about us that we are not aware of. I know it's important to you to do well at your job. I was just hoping that you would see this as an opportunity to do even better. I'll be here late tonight finishing some work, so if you'd like, you can stop by my office before you go home. I want to be sure you are okay.

Mark (*in a condescending tone*): What do you mean? Of course I'm okay.

Laura: That's good. I'll look forward to seeing you later.

Laura felt so weak when she stood up that she was afraid Mark would see how his defensiveness had nearly knocked her off balance.

Back in her office Eric was waiting for her. "So what happened?" he asked. She recounted the whole story. When she finished, the first words out of Eric's mouth were, "I told you so. Aren't you worried about what he might do now?"

"No. Not really. If you think about it, of course he'd react defensively at first. When I acknowledged that his reaction was understandable, he seemed to back off a bit. When I mentioned that we all have blind spots, he just stared at me for a minute. I could see he was processing something. I'm willing to bet that he'll come into my office before he leaves."

"I can't imagine that he will take what you said seriously, much less change his behavior. I'm impressed that you had the courage to speak, but I'm not convinced that you've heard the end of it. Call me later if anything more happens, okay?"

Just before six, Mark knocked on Laura's door.

> Mark: Can I come in?
>
> Laura: Of course.
>
> Mark: Laura, I've been thinking about what you told me earlier, and I must admit, my first reaction was to want to deflect everything you said. The fact is, my wife has been on me for years about the way I talk to people at times. It's an ongoing struggle between us; it always comes up when we're arguing. She likes to use it as a weapon to make me feel bad. She knows it gets to me.
>
> But the way you told me this afternoon felt different. You said you wanted to tell me because you cared about me and our relationship. I could see it wasn't easy for

you, but I could also tell that you wanted to help me. I'm not sure where to go from here, but I just wanted you to know that I'm going to think about what you said. To tell you the truth, I'm not sure I know what to do about it. I may need your help. One last thing— where did you get the guts to come to me and say what you did?

Laura: To tell you the truth, Mark—it's all about dignity; yours and mine. I know we are capable of a whole lot more.

*Option Two: Know a BATNA and Speak Up.* What if Laura had decided that she was dealing with a narcissist and there really was no hope for change? Even if she was right in her assessment that Mark was a narcissist, she could still give him the benefit of the doubt and try giving him the feedback, but before doing so, she needed to do one more thing.

Roger Fisher and William Ury, coauthors of *Getting to Yes,* have developed what, to my mind, is a useful bit of advice for anyone who is about to enter into a negotiation: "Develop your BATNA (best alternative to a negotiated agreement)."[6] What Fisher and Ury mean is that when you enter into a negotiation, you should prepare yourself for the possibility that things may not turn out the way you hoped. In this case, Laura was about to enter into a conversation in which she was hoping that Mark would understand the feedback and change his behavior. Good preparation for failure to meet your primary goal is to develop an alternative that you would be able to live with. That is, Laura should have had a backup plan in mind. She might have decided to transfer to another department if Mark did not take her seriously and change his behavior, or she might have decided to quit altogether. She and her husband had talked about starting their own business, so the tricky conversation could have led to bold, positive steps in that direction. Whatever her BATNA, she needed to be clear in her mind that she had a plan that she could live with.

Going into such a risky conversation without a BATNA left Laura vulnerable. According to Fisher and Ury, we appear much less vulnerable and more confident when we know that we can live with an alternative plan if the conversation doesn't go well. I have seen many people go into job interviews without a BATNA and end up not getting the job because without a backup plan, they unconsciously communicated their neediness and vulnerability. Having a plan enables us to speak without fear of recrimination or consequences.

In this scenario, Laura decided that she couldn't endure Mark's indignities any more, and in spite of believing he had narcissistic tendencies, she still wanted to give him feedback—a part of her didn't feel comfortable writing him off and taking an easy way out. She felt that she had nothing to lose by approaching him, especially now that she had a BATNA—a plan she could live with if the feedback blew up in her face.

In this scenario, Laura's conversation with Mark would look exactly like the conversation she had with him in the first scenario— except that her expectations would be lower, and she would know that she would be all right no matter what happened during or after the conversation.

*Option Three: Say Nothing, but Take Steps to Defend Dignity.* Laura's third option was to say nothing to Mark. She may have assessed the risk of losing her job as too great, and she may have feared she wouldn't be able to find another one that paid as well. She and her husband had just bought a house, and they needed both incomes to cover the mortgage payments.

Her third alternative was to say nothing but to take four steps to defend her dignity in the face of Mark's dignity violations:

1. Avoid personalizing indignities.
2. Reframe what happened.
3. Practice self-restraint.
4. Find social support.

Although others have the power to inflict harm on us by treating us badly, they do not have the power to strip us of our dignity; they may injure it, but they cannot destroy it. It is easy to be confused about this. When someone treats us unfairly, excludes us, or fails to acknowledge and recognize our humanity, it certainly *feels* like an attack on our worth. Yes, a violation feels bad. We are social beings, hardwired to react to the way others treat us, and we are all vulnerable. Although we may not be able to avoid feeling the effects of a violation, what we can do is determine how we are going to make sense of it after it happens. The responsibility is ours to stay anchored in the truth that, as human beings, we are the embodiment of dignity. Once we have this truth clear in our minds, others will find it hard to challenge. The acceptance and recognition of our own worth is the best defense against human vulnerability.

As with all threats to our dignity, challenges leave us vulnerable to thinking that if others treat us badly, then we are unworthy. Allowing others to determine our worth is something we have to consciously fight every day of our lives. Laura had to remind herself that Mark was capable of inflicting harm but that her dignity was sacrosanct—no one could take it away.

Although Laura could not help but feel the impact of Mark's harmful behavior, she could *reframe* the way she thought about it. Kevin Ochsner, a social-neuroscientist at Columbia University, has evidence to show that the way we think about an event can change our emotional experience of it. This kind of cognitive reappraisal affects our neurobiology in a way that decreases the involvement of our old brain (the amygdala) and increases activity in the new part of our brain (the anterior cingulated cortex), which has the power to change how we feel.[7]

Reframing what happened is directly tied to not personalizing the violation. Instead of personalizing Mark's undignified behavior (let's say he made an inappropriate joke about Laura during a staff meeting), she could tell herself that even though she was offended, she would interpret the violation as the result of Mark's lack of social skills—he could have told the joke about anyone: Laura just happened to be the target.

Or, let's say that he did something even more hurtful—he yelled at her at a staff meeting because she didn't agree with a plan he had proposed. Even a good reason to disagree wouldn't justify his aggressive outburst. Although she would feel the impact of the injury, she wouldn't have to take it personally: the incident said more about his behavior than hers, she could decide, and it reflected his inadequacies, not hers.

If Laura personalized the insult and did not reframe the incident as Mark's problem, she could walk away emotionally devastated, believing that he had robbed her of her dignity. Instead, by reframing it, she could walk away feeling as though she had taken a hit but never questioning her worth.

By not personalizing Mark's violations and reframing the event, Laura could keep her dignity intact. She could also do something else: *not take the bait.* She would know that if she had responded to Mark's bad behavior by lashing back, she would have compromised her own dignity by returning the harm. But she would not have let his bad behavior determine how she was going to act.

Although the first three steps (avoid personalizing, reframe, and practice self-restraint) would be essential for Laura in defending her dignity, she would need to take one more step to buffer the impact of ongoing indignities: to *share* what happened with other people.

The research on social support and its positive effects in mediating stressful situations is clear. People who have strong positive relationships with others recover more easily from stressful events than those who do not.[8] Other robust evidence indicates that human beings react to threats not only by fighting and fleeing but also by seeking connections with others. When we experience a violation to our dignity, one way to recover from the toxic effects of the injury is to seek support from others who care about us. When we are in the presence of people with whom we have good, nurturing relationships, there is a biological explanation for why we feel good with them. The hormone oxytocin is released, creating pleasurable feelings. It has the power to turn around the negative effects of dignity violations, it can decrease the stress hormone cortisol, it lowers blood pressure and all sympathetic nervous system activity. In short, it can counter all the reactions that are triggered when we experience a hurtful event.

Good relationships, then, are beneficial to our well-being. But in Laura's case, she and her colleagues were facing repeated dignity violations from Mark, so it was crucial for them to develop a strong support team to reduce the negative effects. They needed to be ready to come to the aid of any one of them who took a hit. One of the worst possible scenarios would be to experience a violation and have no one to turn to.

It is worth noting at this point that one of the most pernicious reactions to dignity violations is the shame response; because of the shame we feel, we don't want to talk about the violation with other people. As we learned earlier from Scheff and Retzinger, and as we all know from experience, it is embarrassing to admit to feeling ashamed; we run the risk of looking bad or weak in the eyes of others. The result of bypassing a feeling of shame can be disastrous, however. Shame has the power to misdirect our moral compass; it can send us down the path of conflict and prompt us to act with righteous justification and perpetrate all kinds of harmful behaviors.

The research on the power of social support shows how critical it is to overcome the temptation to hide or deny our embarrassment. One nurturing conversation in a good, supportive relationship can begin to turn around the toxic effects of shame.

In exercising this third option—to say nothing to Mark about his hurtful behavior but to take subsidiary steps to shore up her dignity—Laura would be making the best of a bad situation. I am not sure that option three is the best one, however. Having researched the effects of toxic relationships on our physical and mental health, I am concerned about long-term exposure to people who are chronically unaware of how they affect others. The four steps in option three that I have suggested are better than doing nothing at all, but in the long term, it may be better to take option two—to have a strong BATNA before speaking up—or even to pick a fourth option: quit.

17

# Being the Victim

*Don't assume that you are the innocent victim in a troubled relationship.
Open yourself to the idea that you might be contributing to the problem.
We need to look at ourselves as others see us.*

We are always ready to ward off incoming information that feels threatening to us: our self-preservation instincts kick in. It is worth repeating that although these instincts most likely evolved originally to protect us from physical threats, we respond today to psychological threats—violations of our dignity—as if our lives were on the line. One obvious example can be seen in the way we instantly assume the position of the victim when a relationship goes wrong. Our internal default setting externalizes the problem even if we contributed to it. The temptation to see the other person as the perpetrator and oneself as the innocent victim is one of the greatest obstacles to resolving conflicts in relationships. Our need to be both right and done wrong by is an outdated survival strategy that creates big problems for us today.

We can change our internal default setting from victim mode to self-questioning mode: "Might I have contributed to the breakdown of this relationship?" Let me describe how.

❁

When my colleagues and I were three days into a workshop at Harvard on Israeli-Palestinian relations, we were wondering, with a fair amount of despair, whether we were doing more harm than good by bringing the two parties together. We could not break the dynamic that was prevalent during the sessions—it appeared to be a competition for victimhood. True, there was an asymmetry of power between the two groups: the Israelis had control of the Occupied Territories, and the Palestinians were subject to their domination, making the Palestinians' sense of being victimized concrete and observable. But the Israelis also felt victimized, which was difficult for the Palestinians to see. Given the parties' mutual failure to empathize with the tragic and painful aspects of the reality of the other, neither side was able to accept the role they had played in the other's suffering. The consciousness of those present was dominated by their own painful experiences, an attitude that the Harvard psychiatrist Judith Herman describes as one of the primary symptoms of emotional trauma.[1]

What is glaringly obvious in troubled relationships is that the normal capacity to empathize vanishes. Someone outside the conflict can, much more easily than either party, see how and why both parties are hurting, and can also perceive the complexity of the dynamics that propel the conflict. For example, when observing friends who each claim to be the innocent victim of the other's bad behavior, you can see that both are contributing to the problem. Although in certain circumstances the line is clearly drawn between victim and perpetrator, the lines are seldom so sharply incised: usually both parties have some role to play in the breakdown of the relationship.

Why, when we are embroiled in conflict, are we unable to see our contribution to it? In *The Age of Empathy*, Frans de Waal provides a biological explanation. He explains that psychological ties hold people together. The capacity to empathize with and care about one another is just as much part of human nature as the capacity to be selfish and concerned only about one's own interests and well-being. De Waal provides examples of the innate desire to connect in the primate world, and he also explains how our own neural circuitry is preprogrammed to feel and empathize with those with whom we are in relationships.[2] In early human history, the capacity to share others' feelings helped bonds to form naturally, creating bands of social groups. People in groups were safer than those who roamed the jungle or savannah alone.

If it is our natural state to want to be in connection with others, if we have an instinctive inclination to empathize with them, what happens to that primal empathy when relationships break down? Why were the Israelis and Palestinians not able to empathize? Why were they not able to feel the pain and suffering they had caused the other side to experience? What had happened to their mirror neurons?

De Waal explains it this way: "Treated with hostility, we show the opposite of empathy."[3] When we experience a threat to our well-being that causes the breakdown of trust, another instinctive reaction replaces primal empathy: the impulse to fight or flee. When we are in defense mode, our limbic system is activated, and the normal control that we have achieved over our emotional reactions is lost.

Under attack, our emotional circuitry (the amygdala) goes on a rampage, and the part of the brain that has the power to override our emotions becomes incapacitated.[4] One of the functions of our emotional brain, especially the amygdala, is to react to potential threats to our well-being. As we have seen, it knows nothing about empathy, problem solving, or taking the perspective of the threatening other. All those other brain functions halt. When the threat persists, which happens in intractable conflicts like that of the Palestinians and Israelis, the self-preservation instincts get stuck in overdrive, and there is little hope for restoring the primal human bond. Is it any wonder that both sides in protracted conflicts see themselves only as victims? The internal mechanisms responsible for the humanizing capacity to feel with the other are on permanent pause.

In the Palestinian-Israeli workshop, with only one session left, the facilitators met to figure out how to proceed. The workshop was one that Herbert Kelman and I had organized as part of a class on the social-psychological dimensions of international conflict. The bonus for the students occurred at the end of the semester, when we invited Palestinians and Israelis for a live encounter. Most of the time the students were behind a two-way mirror observing the dynamics of the workshop (with the knowledge of the participants).

In the facilitation meeting before the final session, a number of students were present, and we asked them if they had any insights to share. Ian Wadley, a quiet and unassuming student from Australia, said: "It seems to me that because each side sees itself as the victim and is unable to let go of that perception, it doesn't make sense to ask either side what it needs the other side to do to prove that the other side is trustworthy. One way to shift these stuck dynamics would be to ask the people on each side to talk about what they themselves could do to prove to the other side that they themselves are trustworthy."

"That's it," everyone realized. How could self-perceived victims trust those they saw as longtime offenders? There wasn't anything the offenders could say about their own actions that would bring instant trust. I looked at Ian Wadley and smiled.

Back in the workshop room we explained what we wanted the two parties to do: talk about what they themselves could do to build mutual trust. People on both sides looked stunned. They were so used to talking about what they wanted from the other side that talking about what they themselves could do required a shift in gear. The question forced them not only to reflect on their own actions but to think about what it was like to be in the others' shoes—what it felt like to be on the receiving end of what they themselves did.

Ian Wadley's brilliant question achieved two things. First, it bypassed the direct question that was next to impossible for the parties to answer: "What are you doing to contribute to the problem?" Instead, it circled around the question, allowing the parties to "tell all the truth but tell it slant."[5] Second, by provoking thought about the situation viewed as a whole, it restored both sides' capacity to empathize with one another.

An Israeli participant was the first to respond. He said, "We can stop the humiliating treatment of Palestinians at the border crossings. It is appalling when an eighteen-year-old solider yells at an elderly member of the Palestinian community who just wants to cross in order to get to work. There is no need for that."

Moments later, a Palestinian spoke up. "We can intervene in our schools to stop the perpetuation of negative stereotypes of Israelis being taught in our history books. We can encourage face-to-face encounters with Israelis so that as many people as possible can get to see their humanity."

The session went on for more than an hour, and everyone on both sides had a chance to speak. What was surprising was how readily the participants responded to the question. It felt to me that this was the kind of conversation they were yearning to have but that no one knew exactly how to go about making happen. I saw the session as a triumph of dignity—both sides were able to take some responsibility for the suffering they created for the other. Hearing the other side acknowledge, albeit indirectly, its role in the conflict and suggest ways to resolve it released each side's exclusive hold on victimhood. The people at the table were able to bring their whole selves back into the discussion and see the complexity of the situation. The discussion could then focus on shades of gray, not the stark blacks and whites that the emotional brain demands.

❀

One last point needs to be made: It is helpful to have a third party present when this kind of victimhood dynamic overcomes people in conflict. Just as you as an observer can more easily see how both of your friends contributed to the breakdown of their relationship than either one of them can individually, a person who is outside a larger struggle can see ways in which both parties contribute to the problem. What is true for interpersonal relationships is true for international ones. Whether the third party is a trained mediator, a facilitator, a therapist, or even a good friend, it is wise to enlist a third party to be present when our primal instincts are running the show. With a little help, we can tap into the power we have within us to make things right and to regain the better part of our humanity when our fight-or-flight evolutionary legacy tempts us to do otherwise.

# Resisting Feedback

Don't resist feedback from others. We often don't know what we don't know. We all have blind spots; we all unconsciously behave in undignified ways. We need to overcome our self-protective instincts and accept constructive criticism. Feedback gives us an opportunity to grow.

Ted (not his real name), who held a significant leadership position in his company, derived much of his sense of self-worth from having such a high level of authority. He had worked hard to reach his position and was understandably proud of his accomplishments. In his view, his relationships with those reporting directly to him were satisfactory, although staff meetings were tense at times. He often had to exert his authority when important decisions had to be made, but for the most part, he felt that he was doing a good job as a manager.

It thus came as shock when he discovered during a facilitated workshop that his staff felt that he sometimes violated their dignity. Before the session, his staff had believed that an open and honest discussion about his violations was off-limits; they feared retribution, maybe even losing their jobs. As you might guess, being unable to address the issue created unspoken tensions between Ted and his staff; no one felt safe to be vulnerable, which meant the staff endured violations of their dignity with no hope of improvements.

Their fears of how Ted would react if they brought up the subject were well founded. Ted's immediate reaction to hearing what he felt was negative criticism was to go into primal-emotional default mode. He spent part of the workshop session defending himself, trying to convince his staff that they hadn't experienced what they said they had experienced. During his tirade, his staff just sat there with hopeless looks on their faces.

What I was seeing was a reenactment of the dynamics that occurred in staff meetings. Instead of trying to understand his staff's experiences, he told them all the reasons they were wrong. He had no curiosity about why they felt the way they did—he was too busy defending himself. There was no acknowledgment or recognition of their perspective. Ted's defensive strategy left him in control—he had the power of his position to do that. But what he failed to see was the effect his behavior had on his staff. They felt that their concerns didn't matter. And they felt resigned to enduring indignities as part of the job. But the indignities created resentment, and resentment accumulates. One day even the slightest provocation could trigger a blowup.

Ironically, the staff felt that in personal matters, Ted was there for them. He was often kind, and he showed compassion for whatever they were going through. But staff meetings were a different story.

Now that the off-limits truth was out, the tension in the room was electric. To keep Ted in the room, I needed to help him see that the feedback from his staff was not addressed to the whole of who he was. It was about a small but significant part of him that was getting him into trouble with people. The feedback was not intended to hurt Ted; rather, his employees were illuminating a blind spot, something we all have. But Ted's self-preservation reactions dominated his response. His knee-jerk reaction was to resist the feedback and defend what felt like a threat to his dignity. His initial response came so fast that we could do little but let the emotional hurricane run its course.

In an earlier session with Ted and his staff, we had talked about the importance of being able to take feedback from others in order to help us see what we cannot see ourselves. At that time, the discussion was in the abstract—there was nothing threatening about agreeing that feedback was invaluable. This session was different: we were dealing with a live example. I was grateful to have the opportunity to put the theory into practice. Taming the evolutionary threats to our dignity is difficult work, but I had faith that Ted would be able to do it and even model the behavior for his staff. In the end, the workshop entailed an excellent demonstration of the innate human resistance to feedback plus a demonstration of how to overcome that resistance.

Ted finally accepted the feedback as feedback. Once he calmed down after his immediate emotional reaction, he was able to see that no one was trying to make him feel bad or look bad. What his employees and I tried to make clear and kept repeating like a mantra was that he was a good person and, like every other human being, he needed help to see his blind spots.

The workshop also provided the perfect opportunity to show the importance of the distinction between one's worthiness, which is inviolable, and one's behavior, which is open to judgment. Disentangling these two concepts is not easy, because our emotional brain (the Me) cannot make the distinction. To our limbic system, feedback feels like a threat to our dignity. We have to rely on another part of ourselves (the I) to see the truth about the matter. It took time for Ted to see that what his staff was offering him was an opportunity to correct something about his behavior that was dysfunctional. A certain behavior was getting him into trouble not only with them but probably with other people and in other relationships.

I would like to examine Ted's reaction from another perspective. Yes, his immediate instinctive response was partly due to something we all share—the deep desire not to want to look bad in the eyes of others and to avoid the feeling of shame that comes along with looking bad. This is our default reaction: to resist feedback. Another way of looking at Ted's reaction is in terms of his understanding of dignity.

❋

Robert Kegan, a world-renowned scholar in the field of human development, has spent his career trying to explain the psychologically rich and complex process by which we come to an understanding of the world and our place in it.[1] He looks at how the human mind makes sense of things, how our internal dialogue changes as we grow and develop, and how the workings of our mind affect our lived experiences. He sees a "continuum of mental complexity." Our thinking about ourselves and the world changes from a distorted egocentric view in childhood to an adult view that more accurately represents reality in all of its complexity.[2]

Using the basic structure of Kegan's elegant theory of how our understanding of ourselves and the world evolves throughout our lives, I will suggest how our sense of worth evolves and changes as well. How we come to know our worth—an aspect of our meaning-making that is emotionally very sensitive—is a fragile process. Our sense of worth is also directly linked to the quality of our relationships, especially those early in our lives that set the stage for our later interactions with others. If, as children, we have repeated experiences of being treated badly, if our dignity is routinely violated, the development of our sense of worth will be compromised, if not arrested. Nothing sets off a primal reaction like a violation of our dignity. It affects our view of the world and upsets our emotional stability. Very few things upset our emotional balance like being treated unfairly or as inferior. To my mind, the only comparable disturbance is brought by loss—the loss of someone or something that we deeply care about. What holds our inner world together at any age—what gives us our integrity, our inner coherence—is our dignity.

Kegan's theory of adult development focuses attention on the cognitive aspects of our growth. He acknowledges that the progression to more complex thinking has an emotional component, especially when it comes to managing the shift from one level of development to another.[3] In changing an aspect of who we are, which is necessary to move to the next level of development, we have to let go of our current way of holding ourselves together. Not only does our thinking have to change, so does the source of our sense of worth. This kind of change is emotionally volatile. Far less is at stake when we learn how to think abstractly or when we learn to take the perspective of others. Both are much less threatening mental events than reevaluating our sense of worth. The development of dignity requires special attention because it is so tender and we are so vulnerable while it develops.

We fear letting go of our current way of defining ourselves, and that fear keeps us from making changes, even changes that we desire. Kegan doesn't use these words, but I believe that what is at the core of our emotional resistance to change is the fear of losing our dignity. If our sense of ourselves, that which holds us together internally, is threatened, we feel anxiety. No wonder we cling to what is known, even if it is not serving us well.

While under emotional attack, we can't comfortably leave ourselves open to change—that would require more vulnerability than we can handle. We are too busy protecting ourselves and our psychological survival —that is what it feels like. How can we navigate through the emotions that throw us off course when the need for change is before us?

When our survival is at stake, whether in the form of a physical or a psychological threat, our emotional brain alerts us that danger is looming. As we know, conflict is a standard way to resolve the threat. But as Kegan points out, conflict has another important function. It can be just what we need to take us to the next level of our development. Drawing on years of research, he claims that "optimal conflict" is what we need to see the limits of our current ways of making meaning; it is what we need to recognize the blind spots that are creating problems in our relationships. By "optimal conflict" he means the "persistent experience of some frustration, dilemma, life puzzle, quandary" that causes us "to feel the limits of our current way of knowing," but "we are neither overwhelmed by the conflict nor able to escape or diffuse it."[4] Optimal conflict, then, is positive, productive conflict.

A lot of good can come from conflict. As Jean Baker Miller argues, we often need to wage "good conflict" because something in our way of making meaning needs to change and because something about the way we relate to others needs to change.[5] In other words, the ways in which we are holding ourselves together are causing problems. If we are constantly in conflict with others, it is time to engage in self-reflection, to see ways in which we are arresting our own development.

If we use conflict constructively, as Miller suggests, we can move to the next developmental level. Instead of trying to preserve our current way of protecting who we think we are and how we derive our dignity, we can open ourselves to the possibility of a new way of anchoring ourselves. The new way could be more inclusive of the reality of the other. It could give us a more complex understanding of why we are in conflict and enable us to see what we could not see but what others (outside observers) could see all along—how our limited perspective is creating pain and suffering.

Now that I have introduced some of Kegan's basic ideas about development, let me use them to explain the emotionally sensitive process of undergoing developmental changes in our understanding of dignity. The development of our understanding of our worth takes place in the context of our relationships. We start out completely dependent on others for dignity, then we shift the balance away from others while we seek an independent assessment of our worth, and finally we arrive at a place where we recognize that our dignity is once again tied to others, but at this more advanced state of awareness it is interdependence that we seek, not dependence. I am proposing, then, three developmental stages of our understanding of dignity: dependence, independence, and interdependence.

*Stage One: Dependence.* What we need from our relationships as children is very different from what we need later. Early in our lives, the content and tone of that inner dialogue about our worth is not only vulnerable to but dependent on others and their treatment of us. If the environment in which we are raised is nurturing and respectful enough, if we have the care and attention we need, we can establish the core of our sense of worth. We experience our dignity because others treat us as worthy.

The danger at this level of understanding is that we remain stuck in the belief that our worthiness is determined exclusively by how others treat us. As good as it feels to be treated well, this level of understanding is situated in our Me, the part of ourselves that seeks praise, status, and approval and compares us to others.

Although this external sense of worthiness begins to develop in childhood, it can remain into adulthood. People who are at this stage feel that their dignity is in the hands of others. If they do not receive praise and recognition, or if they are treated badly, they experience self-doubt, and they expend energy trying to put an end to their feelings of failure and inferiority by looking for or demanding external recognition of their worth.

*Stage Two: Independence.* The next advancement in our understanding of dignity requires a qualitative shift in our experience of ourselves and our relationship to others. In the independence stage, we feel the futility of linking how we feel about ourselves solely to the way others treat us. At this point, we recognize that the most reliable source of our dignity is inside us; our sense of worth becomes internalized. Our inner voice shifts from the Me to the I. We realize, even if we are treated badly, that we are inherently worthy, that our life has unquestionable value, and that we matter. We are strong enough to endure emotional assaults. What makes us vulnerable at this level is not living up to our own ideals.

We are able to make a distinction between our behavior, which is open to scrutiny and critical judgment, and our inherent worth, which is not. At this level of self-awareness we recognize that because we are evolving beings, we all have blind spots, which cause us to make mistakes that can have a negative effect on others and ourselves. We can be judged, sometimes harshly, for our blind spots, but being judged does not make us less worthy.

At the same time, although we still enjoy praise and recognition from others, we are no longer dependent on it. Our sense of worth shifts from being dependent on to being independent of the way others treat us. Our internal anchor has been set. We are internally strong enough to endure assaults to our dignity. We have the emotional capacity to preserve dignity without resorting to fighting or fleeing. The conflict stays inside, where the I and the Me struggle for dominance, and the I most often wins out.

*Stage Three: Interdependence.* At this last and most complex stage of awareness, we come full circle. We realize that we in fact do need others to help us expand our understanding, but not in the needy, dependent way of stage one. In this enlightened stage of interdependence, we realize that our experience of our worth can be deepened with input from others. We recognize that there are aspects of ourselves, blind spots, that other people can see that we cannot, and we welcome feedback. To be fully aware of how we may be unknowingly violating our own dignity and the dignity of others, we need input from others.

Finally and most important, at this stage we recognize that the most elevated experience of dignity is achieved in connection with others, where I's converge to become We.

Continued growth, the ability to become what we are capable of being, is inextricably tied to others. We need others to help us understand who we are and how we affect others. Our awareness of our dignity takes on another dimension as we recognize the necessity of being connected to others to maximize the experience of our own worth. The safety and internal comfort that we achieve through and with others represent dignity at its finest.

Although our awareness of our dignity starts out in relationships and ends there, our experience of ourselves in relationships changes as our development progresses. As Kegan says, development is the product of the management of the ongoing tensions between our human needs for autonomy and for inclusion, for individuation and for integration, for the need to see ourselves and the need to be seen.[6] Ultimately, we develop our awareness of our worth in the context of relationships.

As we develop, as our awareness of our dignity shifts from dependence to independence to interdependence, what also changes is our vulnerability. Because we need others when we are young, our vulnerability is greatest then. The needier we are, the more vulnerable we are. When we move into the later stages of development, knowing that we are worthy even if others treat us badly makes us less vulnerable. Then we come full circle, needing others to help us see what we cannot see. The difference between the first stage and the last stage is that in the last stage we are not experiencing the desperate neediness of a child. Instead, we have attained the joyous realization that what makes us most vulnerable is *not* being connected to others. Safety lies in connection. We need each other to become what we are capable of being.

The stages seem to follow logically and naturally one from the other. Why, then, isn't everyone at the final stage of interdependence? It seems so pleasant.

Kegan's research indicates that adults tend to cluster in the first stage of adult development; a significant portion of their sense of self comes from external sources.[7] What this means is that few of us have internalized our sense of dignity, and even fewer of us realize that we can reach even higher levels of awareness of our worth through our connections with others. What makes it so difficult to reach the higher levels of awareness?

My sense is that the failure to develop goes back to the dialectic of dignity: we all yearn for dignity but are liable to doubt that we have it. Since most of us are dependent on external validation for our sense of dignity, we are wide open to dignity violations. And since most of us not only experienced wounds to our dignity early in life but, because of our dependence and vulnerability, continue to experience wounds, we expect to get hurt again when we approach others. "Be cautious, don't trust others, and don't let your guard down"—that is our self-preservation mantra.

Even if a part of us knows that we don't have to worry so much, the ancient part of our brain that processes hurtful experiences does not want us to take any chances. None of us, after all, wants to leave ourselves exposed to the dreaded possibility of feeling worthless or inadequate.

There we are, stuck in an emotional cul-de-sac, but at least we are not emotionally dead. We cling to our shreds of dignity; we hang on to what we know will make us feel good rather than risk losing it.

Prospect theory addresses how we make decisions when risks are involved. Because of our experiences, a loss of dignity feels worse than a gain feels better.[8] Rather than connecting with another and experiencing dignity through that union (a gain in dignity), we avoid the risk of being violated (loss of dignity) and hold back. We play it safe.

What does it take to shift from that emotional default mode? How can we find our way out of the emotional cul-de-sac and away from the dependence stage of awareness of our dignity? How do we convince our Me to let our I take charge?

First of all, we have to know that we have an I. To do that, we have to accept our own worthiness. Once we recognize our I, our dignity is in our own hands, and we feel safe enough to be vulnerable—safe to leave the protection of the cul-de-sac and venture out on a new road of independence. Our destination is safety in the hearts of others—at a new interdependent level, where our I intersects with others' I's to create a We. At that stage in our understanding of dignity, connection not only makes us safe but brings us back to where we started—in communion with others—but knowing as well as feeling that connection is a good idea.

When Ted finally made himself vulnerable and accepted feedback about the ways he unknowingly violated his staff's dignity, no one was more surprised than he at the positive reaction from those who had offered the feedback. Vulnerability is disarming. It may seem like a paradox that opening up—that is, becoming vulnerable—opens people's hearts, but in developmental terms, it is a paradox only for those in the dependence stage of awareness of their worth. Vulnerability then is perceived as a liability. The inner dialogue of a person at this stage goes something like this:

> If I make myself vulnerable and accept feedback from
> others about the way I am violating their dignity,
> I will look bad and end up feeling bad. What's worse,
> I will not get the praise and recognition that I need
> to feel good about who I am.

It is easy to see the equation of vulnerability with shame and dread.
If my only source of dignity comes from the way I think others see
me, then I am in trouble.

A person whose understanding of dignity is at the independence
stage would say:

> If I make myself vulnerable and accept feedback about
> the way I am violating others' dignity, I will still be a
> good person because I know that my dignity is not up
> for grabs. I may be doing things that I am not aware
> of that hurt people, and that might not feel so good,
> but at the end of the day, I know that I can tolerate
> the feedback because I am certain about my worth.

Here, vulnerability is not dreaded, because the independent person
has a firm sense of personal worthiness. Feedback is tolerable, even
if it is painful.

Finally, at the interdependence stage of understanding dignity, a
person would say:

> If I make myself vulnerable and open myself up to
> others' feedback, I will be grateful to them because even
> though I know my inherent worth, I also know that I
> can understand the impact my behavior has on others
> only if I allow them to tell me. Further, there are times
> when I need to ask for feedback. I know there are limits

to what I can know about myself, and if I want to continue to expand my base of dignity, I need others to
help me see what I cannot see about what I do that
is harming them. I also know that by making myself
vulnerable, I am allowing others to do the same,
deepening the connection we have together.

This person would not equate vulnerability with dread, either, because
of knowing that one's dignity is inviolable. The interdependent person
goes one step further by recognizing that becoming vulnerable with
others opens the door to an even more expansive experience of dignity,
with everyone experiencing it together. This is my understanding of
interconnectedness and its consequence as it affects our felt experience
of dignity: feeling that the whole is greater than its parts.

I have made the case that what keeps us all hovering around the
dependence stage of awareness of our dignity is the fear that we will de-
stabilize our current sense of worthiness and the inner safety and comfort
it provides. It seems better to hold on to the known than to risk opening
up to the possibility of feeling bad about who we are. Self-protection is
understandable, but with the insight gained from an understanding of
human development, we can achieve greater safety than we have now.
To do so we have to give our sense of dignity our full attention instead
expending energy merely pretending that we feel worthy. With eyes wide
open is the only way to begin to see what we are afraid to see.

❋

Back to Ted and his staff. I have told the story as if the changes
that took place between them happened in a neat and tidy way. The
session didn't go that smoothly. Bringing up topics that have been off-
limits, facing blind spots, and giving and receiving feedback require

raw courage—courage and the strength to withstand the immediate desire to run or fight. Offering and taking feedback also require the persistent loving care and attention of facilitators, who by demonstrating how to proceed and communicating through their actions that they are good people, that their worth is not on trial, reassure feedback givers and takers that they can make it to the next step. And even when newly aware people are able to let go of their outdated way of experiencing their worth, they will sometimes backslide. At times the external pressures are too great to withstand, and sometimes their I may not be strong enough to endure an assault on their dignity. They may revert to letting their Me run the show. But with their new knowledge and awareness, and the experience of another way of being, they know how to get back on track.

When Ted first heard the feedback from his staff about ways he was violating their dignity, his sense of dignity was dependent on external sources (he was in the dependence stage). When evidence was put before him that challenged this source of dignity, he went into an emotional tailspin. How could he feel good about himself if the very thing that was holding him together—his belief that he was good at managing—was being challenged? His emotional brain was engaged, and he fought back.

He began telling his staff the many ways they were wrong in their assessment, which is what people do who are embedded in the dependence stage of awareness and derive their feeling of worth from external recognition. But because he exerted his authority in ways that protected his dignity but violated others' dignity, change was required. He didn't want to change, because he feared that if he let go of his current way of being, nothing would be left. In the end, when he realized that his sense of emotional safety was tied to his desire to look good in the eyes of others, he was taken aback.

The dignity model helped him understand that he was clinging to an outdated sense of his dignity—a belief that he needed the recognition of others (his staff) to feel good about himself. The model also helped him see that the conflicts and tensions in his staff meetings were signals that something about his relationships with his staff needed to change. While his tenacious and often childlike Me tried to convince him that he could still think of himself as a good manager even if he denied the validity of their grievances, his ever-strengthening I helped him take a step into his new future. By opening himself up to the feedback of others, he came to derive his sense of dignity in a new way. The experience showed him that he could endure what was once unthinkable: exposing his fear of inadequacy. Before, he had distorted reality rather than be so vulnerable.

Once he was able to see that feedback was not criticism of his worth and that his staff's goal was not to belittle him and do him harm, he was able to absorb the feedback. This was a turning point. With the help of those who could see what he could not see, he came to understand an aspect of himself that he had previously been unaware of.

The overwhelming insight for him was the recognition that the conflicts he was having with his staff signaled a need for change not only in the relationship but in the way he was guarding his dignity. Instead of interpreting negative feedback as threats, now he could see it as an opportunity to expand his understanding of his worth. His relationship with his staff changed. Over time, they came to feel that they could speak their minds without fear of retribution. Their sense of safety, which he had violated in the past by not allowing them to talk about his violations of their dignity, also shifted. Eventually he even encouraged them to speak up. In the end, everyone felt safer to be vulnerable.

# Blaming and Shaming Others to Deflect Your Own Guilt

Don't blame and shame others to deflect your own guilt.
Control the urge to defend yourself by making others look bad.

A survival strategy that protects us from appearing vulnerable and in the wrong is to blame and shame others for our mistakes. Denial goes hand in hand with this reaction. Shaming and blaming others works well only when we have convinced ourselves that we have done nothing wrong. Because denial is a species-level response, we see it taking place at the international level as much as in personal relationships, in Slobodan Milošević's denial of war crimes in the Balkans and in a spouse's denial of an extramarital affair. Human beings resist having their errors and misdeeds exposed.

A big news story broke in 2010 about Charles Rangel, a prominent member of the House of Representatives, who allegedly accepted corporate-sponsored trips to the Caribbean in 2007 and 2008, violating congressional gift rules. Charges of tax evasion and other misconduct were also brought against him. He was eventually found guilty of numerous ethics violations and censured by the House of Representatives. His initial response when interviewed by the press was that he had no idea who had paid for his trips. He claimed that his administrative staff had never told him who had picked up the tab. Similarly, after reports surfaced that Senator John Edwards had fathered a child outside wedlock, he told the world that one of his staffers was the father, not him.

The impulse to blame others for one's transgressions is not limited to politicians, of course. We all have to fight the temptation, and some of us do a better job at it than others. Shifting the blame for our misdeeds is yet another way for us to protect ourselves from looking bad in the eyes of others. Like the other evolutionary traps, it originally evolved to keep us alive but has now gone awry. Whether we act on the feeling or not, most of us know how tempting it is to cover up a wrongdoing. Our evolutionary legacy makes it our default reaction. The instinct is difficult to ignore; no one wants to be exposed, especially people in the public eye.

Falling from grace in the political world is costly, and the reaction by the public that follows an exposure of error is merciless, if not downright self-righteous. We are quick to judge those who fail to summon the moral strength to fight temptation, although we know full well how difficult it is. We show little sympathy for human weakness and vulnerability, especially for those upon whom we have bestowed power and authority. We gasp in shock and shun those who don't live up to our high moral expectations.

Not that we shouldn't expect moral behavior. But the extent to which we seem to relish the exposure of politicians' indiscretions and the extent to which we condemn them for their failures suggests that most of us do so at the expense of owning up to our own human vulnerability. Failure to empathize prevents us from acknowledging how difficult it is to admit to feeling tempted and how difficult it is to fight temptation. As Frans de Waal points out, the failure to empathize with others means that we have severed our identification with them, and the first thing to go is the recognition of our common humanity.[1]

Psychoanalytic theory has a convincing explanation for what happens when we deny the shared aspects of our human identity. The concept of "splitting" was first described by Pierre Janet in the nineteenth century.[2] He describes it as an intrapsychic form of dissociation that takes place in our ego when something happens that is incompatible with the righteous image we have of ourselves or others. Our consciousness tends to "split"—we discard the bad and unwanted aspects, focusing instead on all that is good about the self and the others. Or we discard all that is good, directing our conscious thoughts to all that is bad.

Another intrapsychic process that often goes hand in hand with splitting is the concept of "projection." It is a defense mechanism that is activated when a person's own unacceptable attributes and feelings are repressed and are projected upon someone else.

Both psychological processes, taken to an extreme, are characteristic of what Melanie Klein later described as the paranoid-schizoid position.[3] I am not saying that our self-righteousness in harshly judging those who succumb to temptation falls into the paranoid-schizoid category, but I am suggesting that the tendency to project our own undesirable impulses onto them is an easy route to take.

Let's examine the speed with which the public condemned and judged Charles Rangel and John Edwards for their unethical and immoral acts. In the case of John Edwards, his violation of the marriage vow seemed particularly heinous because it took place when he knew his wife had breast cancer. But if the public's moral outrage was understandable, the degree and intensity of it suggest that splitting and projection might have been taking place in many who were doing the judging.

How many times have we had to fight the temptation to blame others, not wanting to have our moral weaknesses exposed? Even if we successfully hold ourselves to high moral standards, we can at least empathize with the human vulnerability that Edwards (like thousands of others) has shown. According to Peggy Vaughan, author of *The Monogamy Myth* (2003), conservative estimates of the number of adult Americans who engage in extramarital affairs is 60 percent for men and 40 percent for women, and she believes that the actual number for men is closer to 80 percent.[4]

As for Rangel, if the current exposure of misdeeds in the business world that brought on the near collapse of the global economy in 2008 isn't enough evidence of the ubiquitous failure of those in positions of power to resist the temptation to not only blame others for their unethical acts but lie about their wrongdoing, what more do we need? It is very human for us to be tempted and just as human for us to split our wrongdoings from our consciousness and project our badness onto those who get caught. But what is it costing us?

By failing to accept the truth about our shared humanity—that we all have the initial impulse to do whatever it takes not to look bad, whether it is blaming others for our guilt or engaging in other face-saving behaviors, we are perpetuating a myth about what it means to be human.

Feeling the pull of temptation doesn't mean that we are bad people. The evolutionary past lives in us all, like it or not. But instead of engaging in denial, splitting, or projection about such a fundamental part of being human, a truer and more evolved response would be to accept that we all have hardwired impulses and then to educate ourselves about managing and controlling them. However challenging it is to come to terms with our outdated instincts, continuing to pretend that they do not exist is potentially destructive. Failure to bring the truth about the human experience to light means that we give our impulses far too much power. They may trigger our first reaction, but they needn't be our last reaction. We can choose to behave differently.

We already have the ability to disempower these instincts, but we can't do it if we remain in denial. Until we acknowledge that we all share the same human struggle—a struggle that could be made so much easier with a little self-knowledge and self-acceptance—we will continue to suffer.

After acknowledgment, the next step is making a commitment to develop the discipline to control our default reactions. The reactions may guarantee survival; those who fall from grace by following their instincts don't generally die. But the humiliation that Me followers experience when publicly exposed may make them wonder which is worse, humiliation or death.

Safeguarding our dignity when faced with a threatening situation requires us to develop both self-restraint and self-assertion. When we are tempted to blame others, we need to hold ourselves back from acting on that impulse. When we are tempted to avoid confrontation, what serves our dignity is to speak up—to assert ourselves instead of enabling others to harm us unopposed. Either way, the temptation to compromise our own dignity in the service of self-preservation is something we have to fight. Knowing that everyone else is fighting the same internal battle could make it a bit easier to accomplish.

When we are able to talk openly about our instincts, to learn about them and learn how to work with them, we will be able to remove their tight hold on us. Since I developed the dignity model, I have introduced the ten temptations to hundreds of people. Without exception, they experience relief upon learning that having to struggle against temptation doesn't make them bad people. They struggle because they are human beings, and the temptations are part and parcel of their humanity.

But I am quick to add in every workshop that even though our hardwiring isn't our fault, doing something about our responses is our responsibility. We have the inner resources to redirect the potentially harmful energy that these instincts release. Our brains are remarkably capable of building new pathways around old, well-traveled emotional routes.

# Engaging in False Intimacy and Demeaning Gossip

Beware of the tendency to connect by gossiping about others
in a demeaning way. Being critical and judgmental about others
when they are not present is harmful and undignified.
If you want to create intimacy with another, speak the truth
about yourself, about what is happening in your inner world,
and invite the other person to do the same.

Jason was anxious to become acquainted with Megan, who had been hired to head the new development office of the nonprofit organization for which he was the deputy director. There were lots of internal problems in the organization; lack of funding to support their ongoing activities was one. Everyone thought Megan would be just the right person to help the organization get back on its feet financially. In interviews, she had come across as a savvy "people person." Those who had met with her probably also hoped that she would have an influence on the organization's leadership—specifically, the executive director, Cynthia. (I invented the names in this story.)

Cynthia not only failed to generate enough funds to support projects (one of her main tasks) but often alienated her staff. The board of directors had hired her because she had been highly recommended for her ability to build an organization. She had technical skills but very little insight into how to bring out the best in people and create a work environment in which everyone felt valued and recognized for the role she or he played in the organization.

Jason, for one, had suffered a number of her dignity violations. Cynthia had not included him in important decision making that directly affected him, and she had overridden his decisions, seemingly on a whim. On many occasions she had jumped to negative conclusions about his work without asking him to explain his perspective. What irritated him the most, however, was that all in the organization but Cynthia were held accountable for their actions. But because Jason was afraid to confront her for fear of losing a job that he otherwise loved, he put up with her daily indignities. Jason's resentment of the way Cynthia treated him was building up inside him, although he had convinced himself that it didn't matter and that he could rise above it. He was hopeful that Megan would have an influence on Cynthia. He was looking forward to having a conversation with her.

During Megan's first week on the job, Jason invited her to lunch. At one point, the conversation dragged, so he decided to discuss work. The topic: how difficult it was to work with Cynthia. For one thing, he wanted to warn Megan what to expect. For another, he felt that it would help build their relationship if he shared his insights with her.

There is nothing like good dirt to liven up a discussion. He later realized that although he didn't feel that it was safe to talk to Cynthia directly, he was able to release some of his resentment by sharing negative stories about Cynthia with Megan. He enjoyed getting even with her indirectly by spreading word about her hurtful acts; resentment finds refuge in gossip. After lunch, Jason sent Megan an e-mail message expressing pleasure in their get-together, giving her the title of a book he had recommended, and asking her not to repeat his stories about Cynthia. In retrospect, he felt uncomfortable about his tale-telling.

The next day Jason was called into Cynthia's office, only to see his e-mail in her hand. As it turned out, Megan had printed the message so that she could take it with her to the bookstore, and the message had sat in the communal printer for a while. Cynthia had seen her name and read the message. Incensed at being talked about behind her back, especially to a new employee, she fulfilled Jason's worst fear. Cynthia didn't give him a chance to defend himself. All she knew was that Jason had betrayed her, and she fired him on the spot.

Why is it so tempting to talk negatively about people when they are not around, especially people who have done us harm and whom we are afraid to confront? As with the other evolutionary traps that I have described, there is a biological explanation: gossip gave our early ancestors a survival advantage.

The evolutionary psychologist Robin Dunbar makes the case that gossip was a way to efficiently exchange information in a large group and helped maintain group cohesion.[1] Our ancestors needed to be kept abreast of relationships—who did what to whom, who was safe, who was not trustworthy, and who failed to pull their weight. Conveying and updating information about people who are not immediately present can be an effective way of staying on top of group issues. Gossip is a way to monitor changes in the social network made when one is not present. There is nothing wrong with this type of gossip. When the character of the person being gossiped about is at stake, then it becomes undignified.

Telling gossip is also a way for the teller to convey trust in the person being spoken to. By divulging personal information, the teller of the gossip is saying, "I'm relating this to you because I like you and because I know I can trust you." This is the logic of false intimacy. The dynamic created between the two people indulging in gossip might facilitate an alliance, but at the expense of another person. Herein lies the dignity compromise.

Gossip was originally a way to punish those who had taken advantage of the group. No one likes to be talked about behind his or her back. For Cynthia, being gossiped about felt like betrayal. Gossip is a convenient way of making those who harm others look bad when confronting them would be difficult. For those who lack the courage to face an offender, gossip is an easy, though undignified, way to get even. Jason tried to build his relationship with Megan by sharing negative information about Cynthia. He admitted later that telling tales about her felt like a way to punish her for treating him so badly.

In a dignity workshop that Jason attended, I shared the explanation from evolutionary psychology with him. He felt relieved, although he already held himself responsible for his undignified behavior. He had learned a lesson, one that cost him not only his job but his dignity. The explanation did, however, help him to recover from his humiliation.

It was especially helpful to him to learn that feeling tempted to gossip—and feeling tempted in all the other ways enumerated here—was not his fault; the ten temptations are part of our evolutionary inheritance. That said, it is our responsibility to develop an awareness of the harmful impact of succumbing to the temptations—the harm is done both to ourselves and to others—and to keep from taking the lure. Living a dignified life requires both self-knowledge and hard work; we have to learn to override our harmful impulses.

Jason's story is an example of false intimacy because the connection the two people felt over lunch was at the expense of someone else. For real intimacy we have to make ourselves vulnerable by being truthful about our own lives, not others' inadequacies. The lure of gossip is powerful because it gives us a feeling of connection to the one we are gossiping with. It may have evolved to foster group cohesion, but gossip that is demeaning to the people being gossiped about compromises our dignity.

# PART THREE

# HOW TO HEAL RELATIONSHIPS
# WITH DIGNITY

"Can there be any hope of healing, any thought of reconciliation, without an attempt to face the truth?" This is the question that BBC presenter Fergal Keane asked as he opened the three-part television series *Facing the Truth,* in which victims and perpetrators of the conflict in Northern Ireland came together for face-to-face encounters.

As a convener of numerous dialogues between warring parties all over the world, I have asked myself that question many times. What people experience during war often exceeds their capacity—determined by their biological equipment and mental hardwiring—to process those experiences. As T. S. Eliot pointed out, "Human beings cannot bear very much reality." When reality becomes unimaginable, when the truth is unbearable, when loss and suffering become a way of life, how do people heal and move on? How do they face the truth? There is no short answer to the question, but what I have learned from survivors of war is never to underestimate what human beings are capable of.

In the next chapter, you will read about a powerful episode of *Facing the Truth.* It is my account of what happened between two men, one of whom nearly killed the other. At the end of their encounter, they were able to reconcile. I was one of the facilitators of the series encounters, and the experience gave me new insight into what it takes to begin the reconciliation process.

I hesitate to generalize what I learned from my experience. What I feel comfortable saying is that the episodes demonstrated to me that under the right conditions, human beings are capable of extraordinary acts. To quote William James, "Truth is what works."

Before taking part in the television series, I felt that something significant was missing in my understanding of what it takes to reconcile. I sensed that the need to address the impact of living with unspeakable emotional trauma could no longer be ignored. *Facing the Truth* provided me with the opportunity to use the dignity model to encourage and facilitate the restoration of dignity—which I now believe to be an essential part in putting the past to rest. What you are about to read is my understanding of the significant role that dignity played in these two men's encounter and why I believe that they not only started their own healing but gave us all reason to believe that we, too, are capable of more than we think.

# Reconciling with Dignity

> Could a greater miracle take place than for us
> to look through each other's eyes for an instant?
>
> HENRY DAVID THOREAU

Two men in their late fifties sat at a round table shifting in their seats and looking everywhere but at each other. The room was dimly lit except for the lights positioned by the BBC engineer to illuminate the men's faces. Each face was expressionless yet alert, as if the men were readying for the first strike. Ronnie, a former member of the Irish Republican Army (IRA), had served twenty-one years in prison for nearly killing the man who sat opposite him, a British police officer named Malcolm from Southampton, England.

In March 2005, more than thirty years after the near murder, they were meeting again. The two had agreed to take part in a BBC television series, *Facing the Truth,* that brought victims and perpetrators of the conflict in Northern Ireland face-to-face.

I was present because the BBC had invited me to act as one of three facilitators for the encounters. My first reaction was skepticism; the setup sounded perfect for reality TV. When I learned, however, that Archbishop Desmond Tutu was to be one of the facilitators, I accepted its potential value.

The purpose of the television series was to bring healing to a land where human tragedy abounded and the need for reconciliation between Catholics and Protestants was long overdue. And the programs were planned with the utmost care and responsibility. Everyone involved in making the series, from the producers to the cameramen, understood the fragility of the undertaking.[1] The producers went so far as to hire an expert in trauma work from South Africa, Nomfundo Walasa, to be present in case any of the participants needed help during the encounters.

The producers were well aware that although the two sides had reached agreement on many of the political issues in 1998, in the Good Friday Agreement, the suffering caused by years of violence and hatred between the Catholic and Protestant communities in Northern Ireland had never been openly addressed. If there were to be a chance for reconciliation in the war-ravaged country, the painful losses had to be acknowledged. Someone had to open the doors to truth and healing, and the BBC took that bold first step.

Ronnie had shot Malcolm, the British police officer, in England in December 1974, during the height of the Northern Ireland conflict. The Irish Republican Army had taken its bombing campaign to England to bring the war closer to home for the British government. Ronnie and another IRA man had been holed up in an apartment in Southampton for several days. They were waiting for orders to carry out a bombing when they were inadvertently discovered by the British police; the landlord had called the police after visiting the apartment and discovering two men instead of the woman he had rented the place to.

Malcolm, one of the officers called to the scene, had pursued the IRA men after they fled from the apartment. During the chase, Ronnie turned and shot Malcolm. He watched Malcolm fall to the ground and then ran on, avoiding capture, only to have the police arrest him in Northern Ireland a few months later in a routine road check. Just two weeks before he was to be married, he was arrested, and he was sentenced to spend a good part of his life behind bars.

The facilitators—Archbishop Desmond Tutu, Lesley Bilinda, and I—sat at a crescent-shaped table about a foot away from the men. These types of encounters were nothing new for the archbishop. He had facilitated countless discussions with victims and perpetrators of crimes of apartheid in his homeland of South Africa. Lesley Bilinda was asked to join the facilitation team because she had lost her husband in the Rwandan genocide. I was invited because of my years of experience facilitating discussions between warring parties all over the world.

As a conflict-resolution specialist, I had facilitated many dialogues, but nothing could have prepared me for what was to come. What I witnessed during the BBC encounters shifted my understanding of how to heal the wounds of conflict and deepened my awareness of the sanctity of the human heart.

A miraculous reconciliation took place that day between Malcolm and Ronnie. There was another process besides forgiveness at play during the several breathtaking hours the men spent together at that table. Witnessing it changed my understanding of what it takes to put the past to rest when we have suffered painful indignities in our relationships.

Malcolm and Ronnie were in a war, but indignities like those they suffered are not unique to wars. Each of us has been on a battleground. Relationships, no matter what kind, present opportunities to showcase our humanity or our inhumanity. Malcolm and Ronnie's encounter was a beautiful demonstration of the triumph of humanity. Their story needs to be told—not just to honor the sacredness of their experience but also to give us reason to believe that we, too, are capable of the same kind of healing and reconciliation in our own lives.

*The Encounter.* The cameras were ready to roll. All eyes were on Archbishop Tutu, who was sitting between Lesley Bilinda and me, wearing his stunning purple vestment and a large silver crucifix. He leaned forward and, with a smile, welcomed Malcolm and his thirty-five-year-old daughter, who sat with him at the table, and then Ronnie.

The archbishop told the men how courageous they were to agree to take part in the program. He told them that he hoped they would not only begin their own healing process but would, by example, help others with theirs. He said that they would be asked to tell their stories about what happened the night of the shooting and that they could take as long as they wanted. He explained that we, the facilitators, would ask them questions to get them started and would ask questions for clarification as they went along.

Malcolm began. With surprising calmness he commenced his version of what happened the night he was shot by Ronnie.

"Patrol Sergeant Dodds accompanied me that night. He was the passenger in the panda car I was driving. When the call came on the radio, I recognized my colleague's voice. He said he was being chased by two armed men who were firing at him. He told us his location on Westridge Road, and that's as much as we had. I switched the car around as fast as I could. Westridge Road was only four to five hundred yards away from where we were, so we arrived very quickly."

Malcolm told us in dramatic detail all the events that led up to the chase, from when he arrived on the scene to when he fell unconscious to the ground after being shot. "I thought I fell forward but I didn't," he said. "I must have passed out because when I came to, I could see Sergeant Dodds crouched over me." His voice broke and he stopped. Head lowered, with his daughter's hand on his arm, he wept. For the first time in more than thirty years he was talking publicly about what had happened that night. His children had never heard his story. As with many of the victims who took part in the BBC series, he appeared to have kept his grief locked up.

By allowing himself to be vulnerable, in a setting that honored and protected him, he opened the door to his locked-down wounds and revealed his uncompromised humanity. He was beautiful in his grief. Not only did he open his heart, he opened ours. We bore witness to his strength and his willingness to confront his well-guarded trauma. The truth that Malcolm faced was as much his own as the life-altering story he heard when it was Ronnie's turn to speak.

Ronnie told his story next. "I joined the IRA when I was sixteen years of age. And unlike Malcolm, there were no wages, pension, or housing. It was a voluntary thing to join the IRA. And joining the IRA, at that time, was an emotional response to what was happening to our community. For a number of years there was discrimination, sectarianism, something like apartheid in South Africa. Nationalist [Irish] people never felt a part of the state. They felt alienated from the state and we had to assert the rights of the Irish people through the use of violence. And when you join the IRA, it's not a career. In fact, you know there are two things that are going to happen to you: either you end up in a graveyard or in a prison cell."

He described a life-changing incident when he was walking down the street in Belfast one day with another IRA volunteer. "She was carrying a weapon for me and the British army shot her dead. She was shot ten times. She was twenty years of age. And the fact was, the hatred toward the British state was there." Ronnie said that having his comrade shot at his side "reinforced and redetermined" his view "that the only way to resolve the situation was through the use of armed struggle."

At the end of his remarks, he looked at us and said, "I have no regrets about my involvement in the IRA; in fact, I am proud of it."

The whole time Ronnie was telling his story, he maintained a steely resolve; he was clearly determined not to falter under pressure or to admit regret or remorse. His body language sent the message "Don't take me to where I will not go." His face was stern, his jaw was tight, and his eyes were wide with conviction.

At one point, when I asked him whether he had feelings for Malcolm and his family after listening to his story, he snapped at me. "Of course I have feelings for Malcolm. I have feelings for everyone who suffered in this conflict, and especially for Malcolm." And for the first time, he made eye contact with Malcolm. For a few seconds, they held each other's gaze and didn't look away.

I saw this happen, and the archbishop saw it, too. He recognized an opening. He asked, "Malcolm, is there anything you would like to say after hearing Ronnie's story?" Malcolm turned to the archbishop and looked at him for a few seconds. And then we entered the zone of the miraculous.

Malcolm looked back at Ronnie and said, "What I realize now after listening to your story is how difficult it must have been growing up under those conditions. And I believe that if I had grown up under the same circumstances, I would have done the same thing."

I looked over at Ronnie, who appeared stunned. He took a deep breath, put his elbows on the table, and leaned forward toward Malcolm. I watched his face soften and his shoulders drop. The steely resolve disappeared as he awaited Malcolm's next words.

Malcolm continued. "I never had anyone killed alongside me. If that had happened to me, I feel I could kill. I make no bones about that. I am absolutely certain you and some of your colleagues must have felt exactly the same way. What I'd have been prepared to do about it is another matter. I just don't know at this stage."

Ronnie quickly responded, "When you say you don't know what you'd be prepared to do about it, that's understandable. But if you're in a situation where there's no other road to take, no political road to take, no access to the politics, then there's going to be a vacuum, and in that vacuum, violence will erupt."

Malcolm asked, "Do you feel that had you been the age you are at the time, you would take the course you did?"

Ronnie smiled and said, "That's difficult to answer. Like everybody says, you're always wiser in hindsight. And I do regret an armed conflict was necessary. I do not regret that I was involved in the IRA. To be honest, I'm very proud of it."

"How did you feel about the time you spent in prison? Do you feel it was wasted?"

Ronnie chuckled and said, "Of course I feel it was wasted. But the fact of the matter was, I never regarded myself as a criminal. I was a political prisoner. I believed in what I was doing, I had the support of my community. Even though I lost twenty-one years, I felt I was doing something to help our people."

Both men sat silently for a few seconds. Then Malcolm asked, "You have any children?"

"No children."

"Do you have plans to have any?"

"Well, I hope so. I don't know—"

"What I'm really leading up to is the fact that to have been in this situation and to realize that had I been killed that night, my youngest child wouldn't even exist. And I think about what a tremendous loss that would be. I've got three amazing kids and I love them to bits. I feel very awkward having put them through this situation. I said I believed it was a stupid thing to do and I've always held that opinion. Whether I'd do it again under the same circumstances, I'd like to think I would. If I was trying to prevent something I believed was wrong, I'd like to think I'd do it again if the need arose."

"To be honest with you, I'd expect you to," Ronnie said. "I know what you were doing that day was your job. Unfortunately, things happened the way they happened, but I'm glad you had your youngest child. I'm glad you lived. I'm glad you have three beautiful children."

"I hope you have as many at least."

"I hope so. Let's see what happens."

Ronnie expressed a desire to stay in touch with Malcolm; he invited him to come to Belfast and have a talk one day over a pint. He said he wanted to know more about him—what he was doing and what his priorities were.

Near the end of the session, I asked Malcolm, "Do you think it was courageous of Ronnie to come here today?"

"Yes. To sit across the table from the person you almost killed and not be blown away by it is courageous. I have a great deal of respect for him."

We facilitators sat in awe and silence. Finally, the archbishop asked the men how they would like to end the session. They looked at each other for a few seconds, got up from their chairs, reached across the table, and shook hands.

As if what had occurred during the daylong encounter between the two men were not enough, they and their families went into Belfast that night and had dinner together. And they have seen each other since.

*What Happened?* I have wondered what happened between these two men to make their extraordinary reconciliation possible. What enabled them to cross from disconnection to connection? Their reconciliation had nothing to do with forgiveness; it was never asked for or given. But what did happen was equally powerful: they honored each other's dignity and, in so doing, strengthened their own.

In no small way, the environment that was created by the BBC and the facilitation team significantly contributed to their reconciliation. The effect of the presence of a trusted, moral authority—Archbishop Tutu—cannot be understated. His dignity, consistency, and uncommon compassion created the nurturing, nonjudgmental atmosphere necessary for this difficult work. We created a place that set the stage for an encounter with dignity.

In what ways did they honor each other's dignity? First, they both agreed that sitting down together was worthy of their time and attention—that was the initial step. Isn't it much more common to withdraw from those with whom we have been in conflict and refuse to talk to them?

Second, they listened without interrupting or challenging each other's story; they listened to seek understanding. Isn't it much more common for us to listen to our adversaries only to one-up them or to prepare our attack on what they have said?

Third, they acknowledged and recognized what the other had been through. Isn't it much more common to stare without expression at the person we have injured in the heat of a conflict and feel defensive or justified?

Fourth, they honored and acknowledged each other's integrity and, in so doing, created a mutual bond. Once each man heard the other man's experience, he could no longer dehumanize him and exclude him from his own moral community. Our conflict-driven minds create good guys and bad guys, and when we are under conflict's distorting influence, we rarely see ourselves as anything but good. They expanded their understanding by experiencing each other's humanity. As Günter Grass, the Nobel Prize–winning German author and playwright, points out, "Truth exists only in the plural."[2] And because they came to understand each other's reality, the truth they finally uncovered was bigger than their separate stories: they were both victims, caught up in a dysfunctional system crying out for change.

Maybe, with Ronnie and Malcolm's example to inspire us, we can do better at responding to cries for change in our own lives before, rather than after, damage has been done both to ourselves and to those with whom we are in conflict. But we may be a long way away from preventing conflicts altogether. We seem to be more and more likely to enter arguments than to work out our disagreements. What we could use are hopeful ways to put relationships back together again after they break down.

*The Conditions for Reconciliation.* Ronnie and Malcolm showed us one way to put a relationship back together. Identifying exactly what contributed to their reconciliation is difficult, however. Archbishop Tutu, when asked to explain the magic that took place that day, held his hands wide above his head, looked up to the sky, and said, "Thank you, thank you, thank you." I would like to attempt to outline other factors, besides the crucial one of extending dignity, that might have contributed to the positive outcome of Ronnie and Malcolm's encounter as well as the encounters of other participants in the program. These factors are not universal guarantees of reconciliation, but they do reflect the conditions that, to my mind, appeared to contribute to the many positive outcomes that *Facing the Truth* enabled. They are:

1. the need for public acknowledgment
2. the need for safety and nurturing
3. the need for control
4. the need to be vulnerable

As Archbishop Tutu said after the encounters, "There seems to be a need for public vindication that we yearn for when we have been roughed up." When we have been harmed, especially under circumstances that feel unjust, we have a need for public acknowledgment of the pain and suffering that was caused. What struck us all during the encounters was that the perpetrators, with the exception of one, had gone to trial and served time in prison. But that didn't seem to satisfy many of the victims. Trials and jail time, though serving the need for justice, are not sufficient to address the victims' emotional wounds. The victims seemed to have an additional need for a kind of public process that acknowledged their suffering.

Acknowledgment took many forms. Some victims needed to hear from the perpetrators that they were sorry for what they had done, and others wanted to clear up the misunderstandings about what happened. For example, the sister of a Catholic man targeted as a member of the IRA wanted to hear that the British army officer who killed him had made a mistake and shot the wrong man. In another encounter, the widow of a slain man participated because she did not want the perpetrator to have power over her anymore: the acknowledgment that she sought came from within herself. At the end of an agonizing day with the man who had killed her husband, when she walked away no longer fearing or hating his killer, she felt that she had reclaimed a piece of herself, taken from her thirty years earlier when her husband was killed.

The manifest ways the victims needed to have the emotional toll of the deaths acknowledged bespeak the complexity of the human response to violence and loss. What appeared certain for the people who took part in the programs was that the emotional needs related to personal loss were not sufficiently addressed with the signing of a peace accord or the sentencing to jail. For many of the victims, more than thirty years had passed since they lost their loved ones, but judging by their emotional reactions, I would have thought the deaths happened yesterday. Emotional wounds don't automatically go away with the passage of time.

The need for acknowledgment on the part of the perpetrators took yet another form. They seemed to want to tell the story of the undignified and demoralizing conditions under which they had been raised. Many of them talked about living in economically and spiritually impoverished settings—both the IRA men and the Loyalist paramilitaries. They were seeking not forgiveness but an understanding of the conditions that contributed to who they were and why they had decided to engage in violence as a means for change.

When the victims acknowledged how difficult it must have been for the perpetrators to grow up under the conditions they described, the dynamics between them shifted. The acknowledgment enabled them to connect at the human level, restoring the humanity to the relationship.

The moment when the victims acknowledged their perpetrators was always profound. What enabled their uncommon compassion and generosity toward the people who had caused them such suffering is a matter of wonder. We tend to overprotect victims, but one thing I certainly learned was, never underestimate the power of a victim.

In Judith Herman's often-quoted book *Trauma and Recovery,* the author describes three stages of recovery from traumatic loss: establishing safety, reconstructing the trauma story, and restoring the connection between survivors and their community.[3] The process that we designed for *Facing the Truth* took these factors into consideration. The BBC producers went to great lengths to ensure the safety of all who took part. They spent time developing relationships with the participants months before the programs were filmed and were available to them at a moment's notice to answer any questions. By the time of the filming, the participants knew exactly what was scheduled to happen. There was no pressure whatsoever on them to participate. The facilitation team also met with everyone before the filming. We did as much as we could to reassure all the participants that our job was to serve their needs.

Upon reflection, I believe that another factor besides safety contributed to the positive outcomes of the programs. Once the safety factor was established, the participants needed to believe that they could make themselves vulnerable and not have to worry about being traumatized again or shamed. Along with the feeling of being safe, the participants needed to feel nurtured by the facilitation team. For them to feel our compassion, care, and tenderness was critical to their ability to move forward. As the archbishop said, "The participants must feel that they are precious and important and that something irreplaceable would be lost if they were not there." Any harshness or judgment on our part would have destroyed the sanctity of the space they shared. The facilitators' task was to communicate to the participants, usually nonverbally, that we could handle whatever came up during the discussion and that we would be there to nurture and protect them.

In the many discussions that we had as a team before we started the programs, one issue became clear: we were not going to have an agenda for the participants. The only structure we set in place was for them to tell their stories: what happened and how it affected them. If forgiveness emerged spontaneously, wonderful, but we would not try to force victims to forgive if they weren't ready. Because of the lack of control that most victims suffer when a loved one has been killed, taking control away from them by trying to force forgiveness might traumatize them again. We felt that the victims, not external prompters, needed to be in control of the process.

For many of the victims, being in control of what they wanted to say to the perpetrators empowered them. They had come voluntarily to the programs; nothing about the process was forced. When they came, they knew exactly what they wanted to say. There was no loss of words. They were articulate, insightful, and clear. Their exchanges with the perpetrators felt fragile but forceful. The victims all described feeling relieved afterward, as though a burden had been eased.

At the end of each session, the archbishop said to both victim and perpetrator, "Thank you for being vulnerable." The biggest lesson I learned from these encounters is that vulnerability is where the power lies; the magic happens when we expose the truth to ourselves and others and are ultimately set free by it. That is quite a paradox. Our instincts fool us into thinking that deception and cover-up are a good strategy for self-preservation. When our self-protective instincts overpower us—and they can at a moment's notice upon the hint of a threat—our life seems to be on the line. Making ourselves vulnerable at such a time feels like suicide.

At the end of a long day of filming, the archbishop said to me, "Aren't human beings funny creatures? We all do the same thing—we just hate to admit we've done something wrong." He was describing the impulse that stands in the way of reconciling with dignity. If we understand the evolutionary legacy that tries to prevent us from taking responsibility for our actions, if we can circumvent the hardwired shame of admitting that we have done something wrong, we will be so much better at healing ourselves and our relationships than we are now. Fighting the impulse to save face could save our relationships.

The participants exposed the raw material of their grief and human suffering, so what the television-viewing public saw was graphic, uncensored, and disturbing. All the gruesome details of the killings were revealed, along with the torment the killings had caused the victims and their families. But strangely enough, even though most of us on the spot had never experienced such painful losses, we were all affected by them and could all emotionally identify with them. The feeling of loss is universal: the fear of it, the dread of it, the hurt of it. When we can open ourselves to feelings, we are rewarded with the opportunity to connect in the most beautiful way. The walls of separation disintegrate, and we join together in a shared awareness of the painful aspects of our humanity.

*As Good as Forgiveness.* Earlier I made the point that the reconciliation that took place between Ronnie and Malcolm did not involve forgiveness; it was never requested, nor was it given. But something powerful occurred, perhaps best described as the act of honoring dignity.

Part of the magic of honoring dignity is that it quickly becomes reciprocal. Unlike offering forgiveness, which lies completely in the hands of the victim (the victim either forgives or doesn't forgive), honoring dignity engages both the victim and the perpetrator in the healing process; either one can act to shift the dynamics of the relationship. The forgiveness approach and the dignity approach are very different. When dignity is engaged, it is assumed that both parties are in need of understanding—that both contributed to the breakdown of the relationship, that both played a role, though perhaps not an equal role.

When the line between victim and perpetrator is clearly drawn, the process of forgiveness is appropriate. In many conflicts, however, the line that separates victim from perpetrator is not so clear. Although Ronnie was designated the perpetrator for the purposes of the encounter, Malcolm soon understood that Ronnie had been repeatedly victimized by an oppressive political regime. Malcolm's attentive listening enabled him to identify with and empathize with Ronnie. He did not forgive him. He saw no need to.

191

In *No Enemy to Conquer: Forgiveness in an Unforgiving World*, Michael Henderson, the author, describes a similar reconciliation that took place between Joanna Berry, the daughter of Sir Anthony Berry, a member of Parliament, and the IRA volunteer, Pat Magee, who killed Berry during the height of the "Troubles" in Northern Ireland.[4] In private meetings together, they were able to engage with one another much the way Malcolm and Ronnie did. In recounting what happened during their face-to-face discussions, Joanna said that forgiveness was inappropriate in their case.

By carefully listening to Pat's story, she had come to realize the oppressive nature of the broader political context that gave rise to Pat's involvement in the IRA. She, too, said that she might have taken the same course if she had been in his shoes. "Given any situation where I feel my rights have been taken away, I also could make decisions to be violent." She felt that forgiveness would have been condescending, because it would have forced them into an us-them scenario where she had all the power, where she was the one who was right. She said, "Sometimes when I have met with Pat, I've had such a clear understanding of his life that there's nothing to forgive."

In the end, they both saw the world and themselves differently by honoring each other's dignity. Joanna was able to let go of her identity as a victim and see the victim in Pat.

Extending forgiveness or dignity demonstrates a level of emotional development that is the polar opposite of our hardwired reactions of fear, rage, and revenge. In a chapter that I wrote in *Political Culture of Forgiveness and Reconciliation,* a book edited by Leonel Narvaez, I argue that both processes, bestowing forgiveness and honoring dignity, show us what it looks like to not only master our emotional hardwiring but choose to reach out to those who do us harm with a recognition of our shared humanity.[5]

Both offering forgiveness and honoring dignity are very difficult; I am not sure one is any easier than the other. But I am sure that moving ourselves along the continuum from rage to recognition of our shared humanity probably wouldn't happen if it didn't involve the pain of letting go of the wounded part of our identity.

Michael McCullough, a social psychologist, makes the case that the desire to reconcile is part of our evolutionary legacy.[6] It makes sense that we have an instinct within us to reach out to those who have harmed us; our need for each other was as powerful as our need for revenge, and both responses promoted survival. McCullough says that the desire to reconnect with others exists within us, and we can bring it out. Although he describes only forgiveness as a means to that end, honoring dignity is just as good as forgiveness, giving us an option when forgiveness doesn't feel right.

The conditions set in place while making *Facing the Truth*—to allow for acknowledgment, nurturing, personal control, and vulnerability—were all designed to promote and restore human dignity. The focus was on the human dimension and the human cost of the thirty years of violent conflict in Northern Ireland. Although we did not want to discuss the political issues, we did acknowledge the role that politics played in creating the conditions for the conflict—the inequality, discrimination, and unjust policies. In fact, inequality, discrimination, and injustice are violent acts in and of themselves. The injuries that they cause are as damaging as gunshot wounds. For this reason, it was important that the perpetrators described the disempowering and humiliating environment in which they grew up—not to justify their violent behavior but to re-create the environment in which the events of the conflict took place.

We wanted to reveal the human suffering that violent and unjust political environments cause. We wanted to hold it up for full viewing—

every aspect and every angle of it. And we wanted to make it personal. Because the truth of the matter is that the human suffering caused by conflicts is rarely acknowledged and addressed; emotional distress is often ignored, diminished, and even trivialized at the political level. And ironically, it is the unprocessed losses and psychological traumas that maintain the divide between warring communities, even after a peace agreement is signed.

Our goal was to dignify the suffering of the participants by giving it the attention it needed in order to put it to rest. We wanted to give participants a chance to be heard, seen, recognized, and understood. We wanted to give them control by letting them say whatever they wanted to. We wanted to create a sense of possibility for both communities in Northern Ireland—to demonstrate what a healing process looked like so they could imagine a future together, living alongside one another, in dignity.

We wanted to create a process that was humane and nonjudgmental. We did not want to be the arbiters of truth; we wanted to enable it to emerge. We wanted the perpetrators to hear, in the victims' own words, what the loss of their loved ones felt like—the shock, the horror, the disbelief, and the rage—and what it felt like to miss someone so profoundly. We wanted to give the victims who had survived an attack the chance to speak directly to the men who came so close to taking their lives. And we wanted the victims' families, themselves victims, to be able to ask questions of the perpetrators—questions that had haunted them since the deaths of their loved ones.

Besides giving the perpetrators an opportunity to tell their background stories, we wanted to give them an opportunity to come face-to-face with the people whose lives they had so deeply affected so that they could see their victims as normal human beings living with abnormal loss. That is, we wanted to humanize their politicized actions by having them look into the faces of those whose suffering they had caused.

All of the perpetrators said that they had found it necessary to dehumanize their victims in order to carry out the killings. What they failed to point out was that they had dehumanized themselves in the process: they had disconnected from the part of themselves that felt the horror of taking someone's life. In the BBC encounters we wanted to create the conditions for them to feel the effects of their actions. Ironically, the victims were the ones who helped them do that. Having to listen to the victims' heartbreaking stories let them feel the consequences of what they had done. Feeling was an unintended gift from the victims. And when feeling returned, the perpetrators could reconnect with their full humanity—which I believe is what true healing is about. They had a chance to reintegrate all aspects of what it means to be human—the capacity to love and to hate, to connect with others, and to violently disconnect from them.

If people are connected to their own full humanity, they feel the pain they inflict on others as remorse, maybe even shame. And if healing is the goal, then they must reconnect to those painful feelings and integrate them into their self-image. If reconciliation is the goal, both the victim and the perpetrator need to reconnect to each other by feeling each other's loss.

What I experienced in the encounters reassured me that under the right conditions, people who have suffered unspeakable losses in violent conflicts can heal one another and can reconcile with one another. But not everyone is ready for healing and reconciliation even if the right conditions exist. Steps along the way are more difficult for some than for others. The reasons to participate in the BBC series varied from personal and emotional to political. For a few participants, healing was a long way from reconciliation. For two of the victims, clearing their loved ones of involvement in paramilitary organizations was enough. For Malcolm and Ronnie participation brought genuine reconnection.

I would like to think that we all have the capacity to reconcile as they did. When we get into an argument with our colleagues or our loved ones and end up knowingly or unknowingly violating each other's dignity, will we be able to look the other person in the eye and say, "I want to understand you. I want to listen to you and hear your story as much as I want you to hear mine"? I would like to be able to say to my partner that our relationship matters to me as much as my need to be right. I would like to think that one nation could say that to another. Honoring dignity is a way to begin.

# 22

## Dignity's Promise

What are we here for but to make life a little easier for one another?

GEORGE ELIOT

Bruce Perry, the child psychiatrist who also trained as a neuro-scientist, writes that the single most important thing to know about healing from psychological injuries is that loving and supportive relationships have power.[1] He goes so far as to say that although the professional therapeutic relationship is important, what happens outside therapy can be even more healing.

Simple acts of dignity—listening to people and acknowledging their presence, their experiences, and their suffering—can help them recover their self-worth. I have seen it happen many times.

When people suffer an injury to their sense of worth, the antidote is time with people who know how to treat them in a dignified way. As Perry says, these other people don't need to be psychologically insightful or knowledgeable about trauma. They just need to give their love and attention, to be kind and sensitive. What is love if not the act of honoring dignity?

If we do nothing more than be aware of the essential elements of dignity in our everyday lives and practice honoring dignity, we will be making an enormous contribution to the healing of shared injuries.

The effects do not stop with healing. The act of honoring dignity is powerful in and of itself: it makes us feel good and look good, and it brings out the best in each of us. When we extend dignity to others, we open ourselves to the possibility of becoming more caring, more loving, more compassionate—in a word, more *human*. I think I finally understand what Archbishop Tutu means by *ubuntu*. "A person is a person through another person," he believes. "My humanity is caught up, bound up, inextricably, with yours. We can only be human together."[2]

Part of our responsibility for being human is to know who we are and what we are up against in the process of our development. Learning to control our destructive reactions to having our dignity compromised is one of the hardest things we face. At the same time, the joy and well-being that come with the experience of honoring another person's dignity and having our own dignity recognized is one of the greatest emotional highs we are capable of.

George Vaillant, director of the Harvard Study of Adult Development, believes that experiencing the positive emotions of joy, love, compassion, and forgiveness is the essence of a spiritual life. He defines spirituality as "the amalgam of the positive emotions that bind us to other human beings—and to our experience of God." He lists love, hope, joy, forgiveness, compassion, faith, awe, and gratitude as the "spiritually important positive emotions." All of them "involve human connection. None of the eight are 'all about me.'"[3] And the emotions that involve human connection make us bigger, he says; they expand who we are, widening our circle of concern. Living a spiritual life means experiencing joy with one another, feeling the power of each other's worth, and experiencing the human connection that is waiting for us every time we come face-to-face with another person.

I see it as a human imperative that we learn about dignity and the role it plays in our social and emotional well-being. If we remain ignorant, there is little hope of seeing appreciable change in our violence-ridden world. Michel Odent, author of *The Scientification of Love,* urges us to learn how to love one another instead of focusing so much on understanding violence.[4] And what better way to demonstrate our love than to recognize each other's value and worth. Does this give us a clue to what dignity feels like? Does it feel like love?

# Notes

## Introduction

1. Evelin Lindner, "The Concept of Human Dignity," 2006, http://www
.humiliationstudies.org/whoweare/evelin02.php.

2. Evelin Lindner, *Emotion and Conflict* (Westport, CT: Praeger, 2009).

3. Michael J. Sandel, *Justice: What's the Right Thing to Do?* (New York: Farrar,
Straus and Giroux, 2009), 122.

4. Marco Iacoboni, *Mirroring People: The New Science of How We Connect with
Others* (New York: Farrar, Straus and Giroux, 2008).

5. Frans de Waal, *The Age of Empathy: Nature's Lessons for a Kinder Society* (New
York: Harmony Books, 2009); Judith V. Jordon and Linda M. Hartling, "New
Developments in Relational-Cultural Theory," in M. Ballou and L. S. Brown, eds.,
*Rethinking Mental Health and Disorders: Feminist Perspectives* (New York: Guilford
Publications, 2002), 48–70. For ongoing work on relational cultural theory, which
grounds human development in the context of relationships, see Jean Baker Miller's
book, *Toward a New Psychology of Women* (Boston: Beacon Press, 1976), as well as
the work of her colleagues Jordon and Hartling.

6. De Waal, *Age of Empathy;* see Naomi I. Eisenberger and Matthew D. Lieberman, "Why It Hurts to Be Left Out: The Neurocognitive Overlap between Physical
Pain and Social Pain," *Trends in Cognitive Sciences* 8, no. 7 (2004): 294–300.

7. Joseph LeDoux, *The Emotional Brain: The Mysterious Underpinnings of Emo-
tional Life* (New York: Simon and Schuster, 1996). On shame reactions, see Thomas
J. Scheff and Suzanne M. Retzinger, *Emotions and Violence: Shame and Rage in
Destructive Conflicts* (Lexington, MA: Lexington Books, 1991).

8. Jill Bolte Taylor, *My Stroke of Insight: A Brain Scientist's Personal Journey*
(New York: Viking, 2006).

9. Leda Cosmides, John Tooby, and Jerome H. Barkow, "Introduction: Evolutionary Psychology and Conceptual Integration," in *The Adapted Mind: Evolutionary Psychology and the Generation of Culture,* ed. Jerome H. Barkow, Leda Cosmides, and John Tooby (New York: Oxford University Press, 1992), 1–15.

10. Richard Restak, *The New Brain* (Emmaus, PA: Rodale Press, 2003).

11. De Waal, *Age of Empathy;* S. W. Taylor, L. C. Klein, B. P. Lewis, T. L. Gruenewald, R. A. R. Gurung, and J. A. Updegraff, "Biobehavioral Responses to Stress in Females: Tend-and-Befriend, Not Fight-or-Flight," *Psychological Review* 107 (2000): 411–429.

12. Evelin Lindner, *Gender, Humiliation, and Global Security* (Westport, CT: Praeger, 2010); William Ury, *Getting to Peace: Transforming Conflict at Home, at Work, and in the World* (New York: Viking, 1999).

13. Lindner, *Gender, Humiliation, and Global Security,* 6, 7. Evelin Lindner attributes the phrase "security dilemma" to international relations scholar John H. Herz; see J. H. Herz, "Idealist Internationalism and the Security Dilemma," *World Politics* 2 (1950): 157–180.

14. Robert W. Fuller, *Somebodies and Nobodies: Overcoming the Abuse of Rank* (Gabriola Island, Canada: New Societies Publishers, 2003).

15. Daniel Goleman, *Social Intelligence: A New Science of Human Relationships* (New York: Bantam Books, 2006).

16. Scheff and Retzinger, *Emotions and Violence.*

17. Daniel Goleman, *Emotional Intelligence* (New York: Bantam Books, 1995).

18. Felipe Fernández-Armesto, *Humankind: A Brief History* (Oxford: Oxford University Press, 2004), 170.

19. Scheff and Retzinger, *Emotions and Violence.*

20. Michael E. McCullough, *Beyond Revenge: The Evolution of the Forgiveness Instinct* (San Francisco: Jossey-Bass, 2008).

21. N. I. Eisenberger, M. D. Lieberman, and K. D. Williams, "Does Rejection Hurt? An fMRI Study of Social Exclusion," *Science* 302 (2003): 290–292.

22. Scheff and Retzinger, *Emotions and Violence.*

23. Scheff and Retzinger, *Emotions and Violence.*

24. It is a central tenet of human social-cognitive development that we evolve out of a state of egocentrism into one that is more "sociocentric," where we understand that we, others, and the world are inextricably linked.

25. My use of the phrase is inspired by the title of Antonio Damascio's book *The Feeling of What Happens: Body and Emotion in the Making of Consciousness* (San Diego: Harcourt, 1999). On healthy social adjustment, see Jennifer S. Beer, "The

Importance of Emotion-Social Cognition Interactions for Social Functioning," in *Social Neuroscience: Integrating Biological and Psychological Explanations of Social Behavior,* ed. Eddie Harmon-Jones and Piotr Winkielman (New York: Guilford Press, 2007), 15–30.

## PART I. THE TEN ESSENTIAL ELEMENTS OF DIGNITY

1. For more on Burton's and Kelman's approaches, see John Burton, *Conflict: Human Needs Theory* (London: Macmillan, 1990); and Herbert C. Kelman, "Informal Mediation by the Scholar/Practitioner," in E. Weiner, ed., *The Handbook of Interethnic Coexistence* (New York: Continuum, 1998), 310–331.

2. Burton, *Conflict.*

3. For information on the Human Dignity and Humiliation Studies network, see http://www.humiliationstudies.org. Linda Hartling, is the executive director of the network.

4. Evelin Lindner, *Making Enemies: Humiliation and International Conflict.* (Westport, CT: Praeger Security International, 2006).

5. Peter T. Coleman, "Characteristics of Protracted, Intractable Conflict: Toward the Development of a Metaframework—I," *Peace and Conflict: Journal of Peace Psychology* 9, no. 1 (2003): 1–37; Peter T. Coleman, Jennifer S. Goldman, and Katharina Kugler, "Emotional Intractability: Gender, Anger, Aggression and Rumination in Conflict," *International Journal of Conflict Management* 20, no. 2 (2009): 113–131.

6. Lucy Nusseibeh, who lives in East Jerusalem, is the founder-director of Middle East Nonviolence and Democracy (MEND) and director of the Institute of Modern Media at Al-Quds University. She is also a member of the International Governance Council Nonviolent Peaceforce (NP) and cochair of the Awareness Raising Working Group at the Global Partnership for the Prevention of Armed Conflict (GPPAC).

## Chapter 1. Acceptance of Identity

1. Robert Kegan, *The Evolving Self* (Cambridge, MA: Harvard University Press, 1982).

2. Susan Opotow, "Deterring Moral Exclusion," *Journal of Social Issues* 46, no. 1 (1990): 173–182.

3. John Burton, *Conflict: Human Needs Theory* (London: Macmillan, 1990).

4. James Gilligan, *Violence: Reflections on a National Epidemic* (New York: Vintage Books, 1997).

5. The I and the Me were first introduced by William James in his book *The Principles of Psychology* (New York: Henry Holt, 1890).

6. Louis Cozolino, *The Neuroscience of Psychotherapy* (New York: Norton, 2010).

## Chapter 2. Inclusion

1. When I arrived at Harvard, the following people were members of the Program on International Conflict Analysis and Resolution (PICAR): Nalani Ambady, Eileen Babbitt, Ariella Baiery ben Ishay, Cynthia Chataway, Mica Estrada, Maria Hadjipavlou, Susan Korper, Hugh O'Doherty, Win O'Toole, Nadim Rouhana, Pamela Steiner, and William Weisberg.

## Chapter 3. Safety

1. N. I. Eisenberger and M. D. Lieberman, "Why It Hurts to Be Left Out: The Neurocognitive Overlap between Physical Pain and Social Pain," *Trends in Cognitive Sciences* 8, no. 7 (2004): 294–300.

2. Louis Cozolino, *The Neuroscience of Psychotherapy* (New York: Norton, 2010).

3. Judith Herman, *Trauma and Recovery* (New York: Basic Books, 1992).

4. Bruce Perry and Maia Szalavitz, *The Boy Who Was Raised as a Dog: What Traumatized Children Can Teach Us about Loss, Love, and Healing* (New York: Basic Books, 2006).

5. I am grateful to Amanda Curtin, a Cambridge-based therapist who specializes in the healing of childhood trauma, for sharing her insights into the lasting effects of early childhood abuse and neglect.

6. Jennifer Freyd, *Betrayal Trauma: The Logic of Forgetting Childhood Abuse* (Cambridge, MA: Harvard University Press, 1996).

## Chapter 5. Recognition

1. Paul Woodruff, *Reverence: Renewing a Forgotten Virtue* (New York: Oxford University Press, 2001), 41.

2. Richard Dawkins, *The Greatest Show on Earth* (New York: Free Press, 2009).

3. Woodruff, *Reverence*, 4.

### Chapter 7. Benefit of the Doubt

1. Nelson Mandela, *Long Walk to Freedom* (New York: Back Bay Books, 1995), 391; for his press conference where he addressed the issue of white South African fears in the post-apartheid era, see 568.

2. Mandela, *Long Walk to Freedom,* 391.

3. Mandela, *Long Walk to Freedom,* 622.

### Chapter 8. Understanding

1. Personal correspondence with Shulamuth Koenig, founding president of the People's Movement for Human Rights Learning, 1993.

### PART II. THE TEN TEMPTATIONS TO VIOLATE DIGNITY

1. David M. Buss, *Evolutionary Psychology: The New Science of the Mind* (Boston: Pearson, Allyn and Bacon Press, 2004).

2. Jerome Barkow, *Missing the Revolution: Darwinism for Social Scientists* (Oxford: Oxford University Press, 2006), 37.

3. Steven Pinker, *The Blank Slate: The Modern Denial of Human Nature* (New York: Viking Press, 2002).

4. Robin I. M. Dunbar and Louise Barrett, *Handbook of Evolutionary Psychology* (Oxford: Oxford University Press, 2007).

### Chapter 12. Saving Face

1. Steven Pinker, *The Blank Slate: The Modern Denial of Human Nature* (New York: Viking Press, 2002).

2. Daniel Gilbert, *Stumbling on Happiness* (New York: Knopf, 2006).

### Chapter 13. Shirking Responsibility

1. Steven Pinker, *The Blank Slate: The Modern Denial of Human Nature* (New York: Viking Press, 2002).

2. Donna Hicks, "The Role of Identity Reconstruction in Promoting Reconciliation," in *Forgiveness and Reconciliation,* ed. Raymond G. Helmick and Rod-

ney Lawrence Petersen (Philadelphia: Templeton Foundation Press, 2001), 129–149.

3. Robert Kegan, *The Evolving Self* (Cambridge, MA: Harvard University Press, 1982).

4. See Herbert C. Kelman, "Social Psychological Dimensions of International Conflict," in I. W. Zartman and J. L. Rasmussen, eds., *Peacemaking in International Conflict: Methods and Techniques* (Washington, DC: United States Institute of Peace Press, 1997), 191–236; R. Holt and B. Silverstein, "On the Psychology of Enemy Images: Introduction and Overview," *Journal of Social Issues* 45, no. 2 (1989): 1–11; and B. Silverstein and C. Flamenbaum, "Biases in the Perceptions and Cognition of the Actions of Enemies," *Journal of Social Issues* 45, no. 2 (1989): 51–72.

5. Frans de Waal, *The Age of Empathy: Nature's Lessons for a Kinder Society* (New York: Harmony Books, 2009).

6. Kathy Roth-Douquet made a presentation at the conference on the Ivies and the Military, Harvard Divinity School, April 2009. See Kathy Roth-Douquet and Frank Schaeffer, *AWOL: The Unexcused Absence of America's Upper Classes from the Military and How It Hurts Our Country* (New York: Collins Books, 2006).

### Chapter 14. Seeking False Dignity

1. See Bruce Perry and Maia Szalavitz, *The Boy Who Was Raised as a Dog: What Traumatized Children Can Teach Us about Loss, Love and Healing* (New York: Basic Books, 2006); and Donald Winnicott, "The Mirror Role of the Mother and Family in Child Development," in *The Predicament of the Family: A Psycho-analytic Symposium,* ed. Peter Lomas (London: Hogarth Press and the Institute of Psycho-Analysis, 1967), 26–33.

2. I would like to thank Susan Hagan for the insight that "love is attention."

3. Donald Winnicott, *The Child, the Family, and the Outside World* (London: Penguin Books, 1991).

4. Elizabeth Gilbert, *Committed: A Skeptic Makes Peace with Marriage* (New York: Viking, 2010), 5.

### Chapter 15. Seeking False Security

1. Daniel Goleman, *Social Intelligence: The New Science of Human Relationships* (New York: Bantam Books, 2006), 5.

2. Goleman, *Social Intelligence,* 10.

3. John J. Ratey, *A User's Guide to the Brain* (New York: Vintage Books, 2001); Goleman, *Social Intelligence,* 11.

4. Goleman, *Social Intelligence,* 12.

### Chapter 16. Avoiding Conflict

1. Daniel Goleman, *Social Intelligence: The New Science of Human Relationships* (New York: Bantam Books, 2006), 5.

2. Dean Ornish, *Love and Survival* (New York: Harper Perennial, 1998).

3. Goleman, *Social Intelligence,* 5.

4. Alexander Lowen, *Narcissism: Denial of the True Self* (New York: Touchstone Books, 1997).

5. Thomas J. Scheff and Suzanne M. Retzinger, *Emotions and Violence: Shame and Rage in Destructive Conflicts* (Lexington, MA: Lexington Books, 1991).

6. Roger Fisher and William Ury, *Getting to Yes* (New York: Penguin Press, 1982), 97.

7. Kevin Ochsner, "How Thinking Controls Feeling: A Social Cognitive Neuroscience Approach," in *Social Neuroscience,* ed. Eddie Harmon-Jones and Piotr Winkielman (New York: Guilford Press, 2007), 106–131.

8. Ben N. Uchino, Julianne Holt-Lunstad, Darcy Uno, Rebecca Campo, and Maia Reblin, "The Social Neuroscience of Relationships: An Examination of Health Relevant Pathways," and Shelly E. Taylor and Gian C. Gonzaga, "Affiliative Responses to Stress: A Social Neuroscience Model," both in Harmon-Jones and Winkielman, *Social Neuroscience.*

### Chapter 17. Being the Victim

1. Judith Herman, *Trauma and Recovery* (New York: Basic Books, 1997).

2. Frans de Waal, *The Age of Empathy: Nature's Lessons for a Kinder Society* (New York: Harmony Books, 2009).

3. De Waal, *Age of Empathy,* 115.

4. Daniel Goleman, *Social Intelligence: The New Science of Human Relationships* (New York: Bantam Books, 2006).

5. The words are Emily Dickinson's, from her poem "Tell All the Truth but Tell It Slant."

### Chapter 18. Resisting Feedback

1. Robert Kegan, *In Over Our Heads* (Cambridge, MA: Harvard University Press, 1994). It would be impossible to describe Kegan's theory here with the detailed attention it deserves. I urge you to look at his writings for a complete description of his theory of mental development.

2. Robert Kegan, *Evolving Self* (Cambridge, MA: Harvard University Press, 1982).

3. Robert Kegan and Lisa Laskow Lahey, *Immunity to Change: How to Overcome It and Unlock the Potential in Yourself and Your Organization* (Cambridge, MA: Harvard Business Press, 2009).

4. Kegan and Lahey, *Immunity to Change*, 54.

5. Jean Baker Miller, *Toward a New Psychology of Women*, 2nd ed. (Boston: Beacon Press, 1986).

6. Kegan, *In Over Our Heads*.

7. Kegan and Lahey, *Immunity to Change*.

8. Daniel Kahneman and Amos Tversky, "Prospect Theory: An Analysis of Decision under Risk," *Econometrica* 47 (1979): 263–291.

### Chapter 19. Blaming and Shaming Others to Deflect Your Own Guilt

1. Frans de Waal, *The Age of Empathy: Nature's Lessons for a Kinder Society* (New York: Harmony Books, 2009).

2. Pierre Janet, *L'Automatisme psychologique* (Paris: Felix Alcan, 1889).

3. Melanie Klein, "Notes on Some Schizoid Mechanisms," *International Journal of Psychoanalysis* 27, no. 3 (1946): 337–375.

4. Peggy Vaughan, *The Monogamy Myth: A Personal Handbook for Recovering from Affairs*, 3rd ed. (New York: Newmarket Press, 2003).

### Chapter 20. Engaging in False Intimacy and Gossip

1. Louise Barrett, Robin Dunbar, and John Lycett, *Human Evolutionary Psychology* (Princeton, NJ: Princeton University Press, 2002).

## PART III. HOW TO HEAL RELATIONSHIPS WITH DIGNITY
### Chapter 21. Reconciling with Dignity

1. Jeremy Adams from BBC Belfast was the executive producer, John O'Kane and Janette Ballard were co-producers, and Fergal Keene from BBC London was the presenter. The program was filmed at the home of Lord and Lady Dunleath, at Ballywalter Park, Northern Ireland.

2. Günter Grass, from his Nobel Prize speech, December 7, 1999, http://Nobelprize.org/.

3. Judith Herman, *Trauma and Recovery* (New York: Basic Books, 1992).

4. Michael Henderson, *No Enemy to Conquer: Forgiveness in an Unforgiving World* (Waco, TX: Baylor University Press, 2009).

5. Donna Hicks, "Dignity and Forgiveness: Pathways to Emotional Development," in *Political Culture of Forgiveness and Reconciliation,* ed. Leonel Narvaez (Bogotá: Fundación para la Reconciliatión, Bogotá, Colombia, 2010), 99–114.

6. Michael E. McCullough, *Beyond Revenge: The Evolution of the Forgiveness Instinct* (San Francisco: Jossey-Bass, 2008).

### Chapter 22. Dignity's Promise

1. Bruce Perry and Maia Szalavitz, *The Boy Who Was Raised as a Dog: What Traumatized Children Can Teach Us about Loss, Love, and Healing* (New York: Basic Books, 2006).

2. Desmond Tutu, *No Future without Forgiveness* (New York: Doubleday, 2000).

3. George E. Vaillant, *Spiritual Evolution* (New York: Broadway Books, 2008), 5.

4. Michel Odent, *The Scientification of Love* (London: Free Association Books, 2001).

# Selected Bibliography

Barkow, Jerome. *Missing the Revolution: Darwinism for Social Scientists.* Oxford: Oxford University Press, 2006.

Barrett, Louise, Robin Dunbar, and John Lycett. *Human Evolutionary Psychology.* Princeton, NJ: Princeton University Press, 2002.

Beer, Jennifer S. "The Importance of Emotion-Social Cognition Interactions for Social Functioning." In Eddie Harmon-Jones and Piotr Winkielman, eds., *Social Neuroscience: Integrating Biological and Psychological Explanations of Social Behavior,* 15–30. New York: Guilford Press, 2007.

Burton, John. *Conflict: Human Needs Theory.* London: Macmillan, 1990.

Buss, David M. *Evolutionary Psychology: The New Science of the Mind.* Boston: Pearson, Allyn and Bacon Press, 2004.

Coleman, Peter T. "Characteristics of Protracted, Intractable Conflict: Toward the Development of a Metaframework—I." *Peace and Conflict: Journal of Peace Psychology* 9, no. 1 (2003): 1–37.

Coleman, Peter T., Jennifer S. Goldman, and Katharina Kugler. "Emotional Intractability: Gender, Anger, Aggression and Rumination in Conflict." *International Journal of Conflict Management* 20, no. 2 (2009): 113–131.

Cosmides, Leda, John Tooby, and Jerome H. Barkow. "Introduction: Evolutionary Psychology and Conceptual Integration." In Jerome H. Barkow, Leda Cosmides, and John Tooby, eds., *The Adapted Mind: Evolutionary Psychology and the Generation of Culture,* 1–15. New York: Oxford University Press, 1992.

Cozolino, Louis. *The Neuroscience of Psychotherapy.* New York: Norton, 2010.

Damascio, Antonio. *The Feeling of What Happens: Body and Emotion in the Making of Consciousness.* San Diego: Harcourt, 1999.

Dawkins, Richard. *The Greatest Show on Earth.* New York: Free Press, 2009.

de Waal, Frans. *The Age of Empathy: Nature's Lessons for a Kinder Society.* New York: Harmony Books, 2009.

Dunbar, Robin I. M., and Louise Barrett. *Handbook of Evolutionary Psychology.* Oxford: Oxford University Press, 2007.

Eisenberger, Naomi I., and Matthew D. Lieberman. "Why It Hurts to Be Left Out: The Neurocognitive Overlap between Physical Pain and Social Pain." *Trends in Cognitive Sciences* 8, no. 7 (2004): 294–300.

Eisenberger, Naomi I., Matthew D. Lieberman, and Kippling D. Williams. "Does Rejection Hurt? An fMRI Study of Social Exclusion." *Science* 302 (2003): 290–292.

Fernández-Armesto, Felipe. *Humankind: A Brief History.* Oxford: Oxford University Press, 2004.

Fisher, Roger, and William Ury. *Getting to Yes.* New York: Penguin Press, 1982.

Freyd, Jennifer. *Betrayal Trauma: The Logic of Forgetting Childhood Abuse.* Cambridge, MA: Harvard University Press, 1996.

Fuller, Robert W. *Somebodies and Nobodies: Overcoming the Abuse of Rank.* Gabriola Island, Canada: New Societies Publishers, 2003.

Gilbert, Daniel. *Stumbling on Happiness.* New York: Knopf, 2006.

Gilbert, Elizabeth. *Committed: A Skeptic Makes Peace with Marriage.* New York: Viking, 2010.

Gilligan, James. *Violence: Reflections on a National Epidemic.* New York: Vintage Books, 1997.

Goleman, Daniel. *Emotional Intelligence.* New York: Bantam Books, 1995.

———. *Social Intelligence: A New Science of Human Relationships.* New York: Bantam Books, 2006.

Henderson, Michael. *No Enemy to Conquer: Forgiveness in an Unforgiving World.* Waco, TX: Baylor University Press, 2009.

Herman, Judith. *Trauma and Recovery.* New York: Basic Books, 1992.

Hicks, Donna. "Dignity and Forgiveness: Pathways to Emotional Development." In Leonel Narvaez, ed., *Political Culture of Forgiveness and Reconciliation,* 99–114. Bogotá: Fundación para la Reconciliatión, Bogotá, Colombia, 2010.

———. "The Role of Identity Reconstruction in Promoting Reconciliation." In Raymond G. Helmick and Rodney Lawrence Petersen, eds., *Forgiveness and Reconciliation,* 129–149. Philadelphia: Templeton Foundation Press, 2001.

Holt, R., and B. Silverstein. "On the Psychology of Enemy Images: Introduction and Overview," *Journal of Social Issues* 45, no. 2 (1989): 1–11.

Iacoboni, Marco. *Mirroring People: The New Science of How We Connect with Others*. New York: Farrar, Straus and Giroux, 2008.

James, William. *The Principles of Psychology*. New York: Henry Holt, 1890.

Janet, Pierre. *L'Automatisme psychologique*. Paris: Felix Alcan, 1889.

Jordon, Judith V., and Linda M. Hartling. "New Developments in Relational-Cultural Theory." In M. Ballou and L. S. Brown, eds., *Rethinking Mental Health and Disorders: Feminist Perspectives*, 48–70. New York: Guilford Publications, 2002.

Kahneman, Daniel, and Amos Tversky. "Prospect Theory: An Analysis of Decision under Risk." *Econometrica* 47 (1979): 263–291.

Kegan, Robert. *The Evolving Self*. Cambridge, MA: Harvard University Press, 1982.

———. *In Over Our Heads*. Cambridge, MA: Harvard University Press, 1994.

Kegan, Robert, and Lisa Laskow Lahey. *Immunity to Change: How to Overcome It and Unlock the Potential in Yourself and Your Organization*. Cambridge, MA: Harvard Business Press, 2009.

Kelman, Herbert C. "Informal Mediation by the Scholar/Practitioner." In E. Weiner, ed., *The Handbook of Interethnic Coexistence*, 310–331. New York: Continuum, 1998.

———. "Social Psychological Dimensions of International Conflict." In I. W. Zartman and J. L. Rasmussen, eds., *Peacemaking in International Conflict: Methods and Techniques*, 191–236. Washington, DC: United States Institute of Peace Press, 1997.

Klein, Melanie. "Notes on Some Schizoid Mechanisms." *International Journal of Psychoanalysis* 27, no. 3 (1946): 337–375.

LeDoux, Joseph. *The Emotional Brain: The Mysterious Underpinnings of Emotional Life*. New York: Simon and Schuster, 1996.

Lindner, Evelin. "The Concept of Human Dignity," 2006 www.humiliation studies.org/whoweare/evelin02.php.

———. *Emotion and Conflict*. Westport, CT: Praeger, 2009.

———. *Gender, Humiliation, and Global Security*. Westport, CT: Praeger, 2010.

———. *Making Enemies: Humiliation and International Conflict*. Westport, CT: Praeger Security International, 2006.

Lowen, Alexander. *Narcissism: Denial of the True Self*. New York: Touchstone Books, 1997.

Mandela, Nelson. *Long Walk to Freedom.* New York: Back Bay Books, 1995.

McCullough, Michael E. *Beyond Revenge: The Evolution of the Forgiveness Instinct.* San Francisco: Jossey-Bass, 2008.

Miller, Jean Baker. *Toward a New Psychology of Women.* 2nd ed. Boston, MA: Beacon Press, 1976.

Ochsner, Kevin. "How Thinking Controls Feeling: A Social Cognitive Neuroscience Approach." In Eddie Harmon-Jones and Piotr Winkielman, eds., *Social Neuroscience,* 106–133. New York: Guilford Press, 2007.

Odent, Michel. *The Scientification of Love.* London: Free Association Books, 2001.

Opotow, Susan. "Deterring Moral Exclusion." *Journal of Social Issues* 46, no. 1 (1990): 173–182.

Ornish, Dean. *Love and Survival.* New York: Harper Perennial, 1998.

Perry, Bruce and Maia Szalavitz. *The Boy Who Was Raised as a Dog: What Traumatized Children Can Teach Us about Loss, Love, and Healing.* New York: Basic Books, 2006.

Ratey, John J. *A User's Guide to the Brain.* New York: Vintage Books, 2001.

Restak, Richard. *The New Brain.* Emmaus, PA: Rodale Press, 2003.

Sandel, Michael J. *Justice: What's the Right Thing to Do?* New York: Farrar, Straus and Giroux, 2009.

Scheff, Thomas J., and Suzanne M. Retzinger. *Emotions and Violence: Shame and Rage in Destructive Conflicts.* Lexington, MA: Lexington Books, 1991.

Silverstein, B., and C. Flamenbaum. "Biases in the Perceptions and Cognition of the Actions of Enemies." *Journal of Social Issues* 45, no. 2 (1989): 51–72.

Taylor, Jill Bolte. *My Stroke of Insight: A Brain Scientist's Personal Journey.* New York: Viking, 2006.

Taylor, S. W., L. C. Klein, B. P. Lewis, T. L. Gruenewald, R. A. R. Gurung, and J. A. Updegraff. "Biobehavioral Responses to Stress in Females: Tend-and-Befriend, Not Fight-or-Flight." *Psychological Review* 107 (2000): 411–429.

Taylor, Shelly E., and Gian C. Gonzaga. "Affiliative Responses to Stress: A Social Neuroscience Model." In Eddie Harmon-Jones and Piotr Winkielman, eds., *Social Neuroscience.* New York: Guilford Press, 2007.

Tutu, Desmond. *No Future without Forgiveness.* New York: Doubleday, 2000.

Uchino, Ben N., Julianne Holt-Lunstad, Darcy Uno, Rebecca Campo, and Maia Reblin. "The Social Neuroscience of Relationships: An Examination of Health Relevant Pathways." In Eddie Harmon-Jones and Piotr Winkielman, eds., *Social Neuroscience.* New York: Guilford Press, 2007.

Ury, William. *Getting to Peace: Transforming Conflict at Home, at Work, and in the World.* New York: Viking, 1999.

Vaillant, George E. *Spiritual Evolution.* New York: Broadway Books, 2008.

Vaughan, Peggy. *The Monogamy Myth: A Personal Handbook for Recovering from Affairs.* 3rd ed. New York: Newmarket Press, 2003.

Winnicott, Donald. *The Child, the Family, and the Outside World.* London: Penguin Books, 1991.

———. "The Mirror Role of the Mother and Family in Child Development." In Peter Lomas, ed., *The Predicament of the Family: A Psycho-analytic Symposium,* 26–33. London: Hogarth Press and the Institute of Psycho-Analysis, 1967.

Woodruff, Paul. *Reverence: Renewing a Forgotten Virtue.* New York: Oxford University Press, 2001.

# Index